"Jill Weber's incisive description of stress-induced brain fog will resonate with many readers, all the more so in the wake of the chronic stress evoked by the COVID-19 pandemic. To help us cope with and bounce back from this debilitating and demoralizing condition, she provides practical, comprehensive, evidence-based suggestions for tending to our mental and physical well-being and getting more joy from our days."

—**David A. F. Haaga, PhD**, professor of psychology, and director of clinical training at American University

"Jill Weber has written a very important book at a critical time—a time when many are crippled by stress-induced brain fog and fatigue. In this thoughtful, insightful, and research-driven book, Weber goes beyond explaining the brain fog epidemic and offers specific strategies for immediate relief. This book is an invaluable, modern resource for clinicians and laypeople alike."

—**Tanie Miller Kabala, PhD**, licensed psychologist, author of *The Weight Loss Surgery Coping Companion*, and creator of *Rest and Restore*

"In this wonderful book, Jill Weber demystifies the phenomenon of stress-induced brain fog. Her relatable examples and thought-provoking exercises help to illustrate how stress might be impacting one's ability to think clearly. Weber is an insightful guide—helping to restore a sense of focus and well-being. Before slogging through another day, I highly recommend reading this book to feel more control of your mind and your life."

—**Emily Aron, MD**, child, adolescent, and
adult psychiatrist; and associate professor at
Georgetown University School of Medicine

"With more and more people suffering from the cognitive effects of stress, Jill Weber offers a beacon of hope, with actionable strategies to reclaim mental clarity."

—**Andrea Bonior, PhD**, licensed clinical psychologist,
and author of *Detox Your Thoughts*

OVERCOMING STRESS-INDUCED BRAIN FOG

10 Simple Ways to
Find Focus,
Improve Memory
& Feel Grounded

JILL WEBER, PhD

New Harbinger Publications, Inc.

Publisher's Note

NEW HARBINGER PUBLICATIONS is a registered trademark of New Harbinger Publications, Inc.

New Harbinger Publications is an employee-owned company.

New Harbinger Publications, Inc.
5674 Shattuck Avenue
Oakland, CA 94609
www.newharbinger.com

Cover design by Amy Daniel
Acquired by Jess O'Brien
Edited by Brady Kahn

Library of Congress Cataloging-in-Publication Data

Names: Weber, Jill P., 1973- author.
Title: Overcoming stress-induced brain fog / Jill Weber.
Description: Oakland, CA : New Harbinger Publications, Inc., [2022] | Includes bibliographical references.
Identifiers: LCCN 2022018696 | ISBN 9781684039944 (trade paperback)
Subjects: LCSH: Stress (Psychology) | Cognition. | Anxiety. | Mindfulness (Psychology) | BISAC: SELF-HELP / Self-Management / Stress Management | PSYCHOLOGY / Cognitive Psychology & Cognition
Classification: LCC BF575.S75 W436 2022 | DDC 155.9/042--dc23/eng/20220425
LC record available at https://lccn.loc.gov/2022018696

Printed in the United States of America

24   23   22

10   9   8   7   6   5   4   3   2   1   First Printing

With love and gratitude for the stars that light my way,
Matt, Maddie, and Willie.

CONTENTS

INTRODUCTION

Have you ever felt that you can't quite relax? Even when you have downtime, between your tasks, job, schoolwork, children, and other responsibilities, you're not at peace. Instead, your brain defaults to a keyed-up, disjointed sense of yourself—a kind of mental frazzle. Or, do you feel sluggish or struggle to focus, concentrate, and remember what you should be doing, and even as you make this effort, your mind drifts: *Why did I waste the day again? I should be organizing my garage, cleaning out my closets. Why am I so unmotivated?* or *I forget what I was supposed to be doing,* or *I can never get anything done. What is wrong with me?* or *I can't believe how much I used to be able to do and now I am barely holding on.* And before you know it, you've lost another hour to these coiling thoughts. You feel empty and criticize yourself for wasting the opportunity to start that work project, teach your children something new, or do something just for the joy of it.

You are not alone; many people are feeling this at some level. We are living in a changing, uncertain world. We shoulder the burden of too much responsibility and overcommitment in all sorts of domains: work, school, relationships, and family. We are daunted by the world's ongoing stressors: bitter national politics; droughts, fires, and bigger and more frequent storms spinning out of climate change; evolving norms in financial markets; the demands of new technology; the thought of jobs being lost to artificial intelligence; fears of unemployment; global crises like the COVID-19 pandemic; and more.

The absolute immediacy of cell phones may leave us feeling as though we need to be accessible and responsive twenty-four hours a day. To be productive and keep up with the social pace, we check and check again our texts, emails, dating apps, phone calls, breaking news, and social media updates.

These forces affect some of us more than others. But, taken together, they inject a load of stress into our society. Some of my clients explain that they are left with an ongoing sense of "dread," "walking on eggshells," or "waiting for the other shoe to drop." When you layer onto this already anxious reality an additional life stressor, such as a medical illness for yourself or a family member, a divorce, a breakup, a cheating spouse, a death in the family, job loss, or a new baby, you literally stop being able to think straight. This is a state of stress-induced brain fog.

Stressed-induced brain fog is when you're no longer able to focus and remember at your usual capacity. You feel sluggish, scattered, and disorganized. You check boxes and go through the motions of life, but you also feel a lack of meaning and connection with yourself, your loved ones, and your job or other responsibilities.

When stress is chronic, as is often the case with brain fog, our brain's physiological reaction, the fight-or-flight response, remains activated. As a result, stress hormones, most notably cortisol and adrenaline, persistently pulse through the body. Feeling depressed, losing focus, feeling on edge, not being able to remember things, and having trouble concentrating or sleeping can all be linked to chronically high cortisol levels. So, when you feel as if what you're experiencing is more than a bad week or month, you're right.

Many believe that to be successful, they have to be stressed out, overbooked, and constantly plugged in. In fact, some tell themselves that living on five hours of sleep, drinking coffee all day, and never missing an email is precisely why they keep up and remain effective. But these patterns aren't sustainable. Make no mistake, everyone has a breaking point. Perhaps you've already hit yours, and you know it. Or maybe you feel brain fog setting in and you want to know what to do about it before it gets any worse. Whatever your particular situation, living with brain fog is not necessary for success—and adopting the skills in this book will stop your situation from getting any worse.

Stressed-induced brain fog is different from the mental fog that is caused by medical and neurological conditions, such as Alzheimer's or dementia, or the neurological symptoms that may result from surgery, pregnancy, or COVID. This is because stressed-induced brain fog is not the result of a medical condition but a result of stress in your environment and your reaction to it. Over time, an accumulation of life stress leaves you feeling depleted and unfocused, and at some point, the pot boils over and you can no longer healthfully manage.

But there are strategies that can help you change course and bring you out of the fog and into the clear, bright light of day. Just recognizing that what you're feeling is the result of stress and that you can learn better ways to deal with that stress will set you on your way. Each chapter in this book addresses a different aspect of how your mind functions and what it needs to work well; for many trying to pull out of brain fog, understanding these mechanics is motivating and inspires change. These chapters also target each of the key symptoms of stressed-induced brain fog and offer a complementary solution for each symptom. For instance, to overcome the symptom of being spaced out, you'll learn skills that help you cultivate awareness, and to overcome the sense of helplessness that marks stressed-induced brain fog, you'll learn skills to help you develop mastery in your life. The solutions outlined in this book are all simple to implement and evidence based, meaning they come from therapy modalities like cognitive behavioral therapy (CBT) and acceptance and commitment therapy (ACT), mindfulness practices, and research in neuroscience, and research has proven their effectiveness.

SOLUTION 1

Overcome Being Spaced Out by Using Awareness

Sustaining perpetual motion, relentlessly going from one responsibility or task to the next, leaves us spaced out and unfocused. This is because the more overstretched we become, the less mental bandwidth we have to focus and reflect. It's like having too many tabs open on your web browser and not enough processing speed—your computer freezes up. If you have brain fog, then most of your energy is going into maintaining the open tabs in your life, your responsibilities and commitments, leaving you with little free space to figure out what to tweak or do differently.

Obsessively checking boxes without awareness leaves you feeling unsatisfied with your life and with the person you've become. This is because when we're in constant motion, we tend to mentally detach as a way to give our brains a break. You may find yourself going from task to task, responsibility to responsibility, commitment to commitment, but without any real connection to what you're doing. Over time, a pattern becomes established where you robotically perform—taking care of things, kids, parents, work, life—but with no real sense of meaning or pleasure.

This dulled-out, unfocused, detached way of existing is brain fog. You're mentally elsewhere; you're not present in the here and now. Feeling all at once overwhelmed and disconnected means your memory isn't encoding key details, and so you've become fuzzy and a bit spacy. As such, you struggle to remember things, and you feel chronically stressed out and unorganized. At times you likely feel defeated, like you're failing, and that you will never be in control of your life. To take the edge off feeling deeply unfulfilled, you give yourself a break and numb out: through excessive sleep, alcohol, drugs, binge-watching TV, overeating, smoking, pornography... Then, you wake up the next day and repeat the pattern all over again.

Natasha's Story

Take the example of Natasha. A thirty-eight-year-old married mother of three and a part-time accountant, Natasha came to therapy because she wasn't sleeping well and felt constantly stressed. Natasha's thoughts were obsessively focused on her responsibilities and daily duties. She had a never-ending list that she recited during the first few sessions—managing her kids' schedules, school events, work tasks, dinner, laundry, social commitments, chores around the house. I began to feel tired and anxious for her.

By the time Natasha's kids were finally in bed, she would say good night to her husband and then binge-watch TV and eat. This was her way of coping. The next day she'd wake up and repeat the cycle.

Natasha described stacks of random stuff in her kitchen: forms for kids' events, permission slips, paperwork for her job, piles of items she still needed to return, leftover decorations in boxes from the previous Christmas. Mealtime was particularly stressful, as she often had to order in or make a last-minute plan for dinner. She found herself working late and then running out to the grocery store at seven o'clock and cooking at eight. Natasha's fellow mom friends and colleagues would often remind her of an upcoming event or task, and she would act as if she'd remembered and would quickly get herself or her kids ready—but in reality, she was not remembering. On the outside, Natasha pretended to have it all together. On the inside, she felt continually behind and inadequate.

By the end of the first few sessions, I suggested to Natasha that although she had a supportive husband, well-adjusted kids, and generally what you might perceive as a great life, for some reason she wasn't able to enjoy or connect with the good. For the first time in a long time, Natasha cried. She began to recognize how lonely and disconnected she felt from herself, her life, and her happiness.

If you're in Natasha's position, how do you begin to turn this pattern of disconnection and loneliness around? You absolutely can—and it starts by recognizing your own state of mind and the behaviors that are contributing to it.

Recognizing That You're Treading Water

People with brain fog tend to tread water and bob for breath as best they can. All the while the tide is rising and the waves are growing. They are so busy and exhausted, they don't have the mental space to slow down and self-reflect.

Granted, pausing and self-reflecting may feel like just another burden to an already heavy load, as it takes a special effort. Certainly in life, there are circumstances in which you don't have much choice but to go through the motions, at least in the short term. After all, as long as you can keep your head above water, you will survive. However, in the long run, not pausing to become aware of your situation means you cannot grow and reach the stability of dry land. And it means you repeat self-defeating patterns day after day—even though there's often a way out, a way to a more secure life where there's room for increased peace of mind, creativity, greater success, and new opportunities.

You may be so busy treading water that you don't recognize you've been struggling with brain fog. Brain fog absolutely can lift, and the first step is to slow down and build awareness for the patterns of behavior that are working against you. Giving yourself the opportunity to feel better, become more successful, become more present, and develop more meaning depends on first acknowledging your state of mind and how you're functioning or not functioning.

EXERCISE: Understand Your Brain Fog

Most everyone has moments of brain fog, and few with brain fog have all of these symptoms all the time, but consider if these statements describe your general experience of yourself.

You're in the midst of a conversation and then suddenly forget what you were saying.

You're perpetually busy and at the same time feel as if you're always behind.

You forget things and then remember at the last minute and scramble to make it work.

It's difficult to push everything to the side and focus and concentrate on one task.

You feel emotionally numb as if you don't really have access to your feelings or know what you feel about important aspects of your life.

You're "doing life," so to speak, but you're not really *in* your life and in control.

You're busy and distracted most of the day, but then when that's over, you hit an emotional slump, feel irritable or blah.

If you're not busy, you mindlessly scroll on your phone—a never-ending scroll where you lose track of time or feel like you're in a trance.

You feel defeated, as if you will never be on top of things and competent.

You're perpetually exhausted.

In meetings, you tend to zone out, space out, daydream, and lose track of the discussion and can't or don't contribute.

You feel like an imposter, as if you're faking it and you're afraid people will see the real you.

You go through the motions of life but feel fragile inside—like you are about to break.

Some of your friendships have ended, or you're less close with others because of your stress level.

You experience dysfunctional dynamics with your romantic partner or family.

You have missed out on job promotions or lost jobs because of your brain fog.

You are unemployed because of brain fog.

You have medical issues that in part might be because of your stress level, for example, difficulty sleeping, over- or undereating, or high blood pressure.

As you reviewed this list and noticed some familiar experiences, you're likely becoming more aware of how brain fog is affecting your life. Besides feeling disorganized and scattered, you may have less connected and healthy relationships than you could have with those you love, feel less functional in your professional life, or have diminished memory and organizational skills. This book will help you address these and other issues.

Brain fog has many ways of showing up in your life. But what causes brain fog? The reality is that stress, when experienced repeatedly over time, becomes chronic. And this chronic stress impacts not just your body but your brain.

How Chronic Stress Impacts the Brain

Fight-or-flight is an evolutionally wired-in survival mechanism. When you perceive danger in your environment, the amygdala, a small almond structure in your brain that is responsible for emotional processing, sends an immediate message to the hypothalamus. The hypothalamus triggers the sympathetic nervous system to mobilize and manage the threat. As a result of this signal, your body starts to change physically. Breathing hastens; heart rate, pulse, and your blood pressure all increase; and your digestion slows. Your brain has one goal and one goal only, and that is to put all energy into surviving the perceived threat or danger at hand.

When we experience the typical types of acute stressors—say, a spider on the wall that startles us a bit or a car coming toward us that leaves us a few seconds to leap out of the way—our fight-or-flight response kicks in. But then our bodies return to a baseline through the work of the parasympathetic nervous system, and we feel like our natural selves again. Certain stressors, however, are less finite in nature and more abstract than concrete. These kinds of stressors include big life events and crises, such as medical illnesses, job adjustments, a death in the family, or divorce; stressors involved in parenting; or stressors related to our broader social and political climate. Stressors that involve more prolonged suffering, which is often the case with brain fog, can adversely impact the brain. This is because when we experience chronic stress, the fight-or-flight response remains switched on. The stress hormones, most notably cortisol and adrenaline, continue to pulse through the body.

Hormones and neurotransmitters involved in the body's stress response impact our ability to focus, attend, learn new information, and remember. For example, research shows that the neurotransmitters involved in the body's fight-or-flight response impact brain structures involved in learning and memory, including the hippocampus (Vogel and Schwabe 2016). Stress takes a toll on the brain and specifically impairs memory and new learning.

Feeling depressed or low in mood, losing focus, not being able to remember things, having trouble concentrating or trouble sleeping—these can all be linked to chronically high cortisol levels. So when you feel like you are in a fog, you are in a fog. Your brain is not processing your environment in an effective manner. Your ability to remember, effectively solve problems, learn, and take in new information is compromised.

The good news is you can change this cycle and start living the life you want.

Natasha Builds Her Self-Awareness

Natasha came to therapy feeling awful about herself. She was barely keeping up with her many responsibilities, and all the while she felt ineffective. Together, we talked through the mechanics of the body's stress response, and over the next week Natasha tracked her physical symptoms so she could begin to understand how her body's stress response was affecting her. She kept a note in her phone in which she logged her eating patterns, sleep cycles, energy level, physical exercise, and memory. By her next appointment, Natasha's self-awareness had grown exponentially.

Natasha recognized she was only getting five hours of sleep on average, was mostly eating carbohydrates, and when she wasn't busy, felt entirely exhausted. She became aware that during the day, no matter what she was doing, her heart was beating fast and her thoughts would race with lists, mental planning, or other things she should be doing. Natasha reported that while working she would sit at her desk all day, barely getting up other than to use the bathroom. She also realized that she was regularly forgetting things. When she received an email with a project for work or a school commitment for her kids, she was quickly distracted from it, so she wasn't making a conscious effort to remember anything beyond what was going on in that particular moment.

With this new awareness, Natasha started to deeply take in how she was living and functioning. She became aware that she was in a vicious cycle of not giving herself what she needed to be effective in her life—regular sleep, solid nutrition, exercise, and relaxation—while tending to beat herself up for not being more competent. She saw how this self-critical thinking only perpetuated her body's stress response. Natasha stopped punishing herself and instead looked at her situation as brain fog. This awareness motivated Natasha to start making some changes: adjusting her routines as best she could to take better care of herself and to reduce the amount of brain fog she felt; and becoming more compassionate with herself when she was struggling, decreasing her self-blame, so she could focus her attention on what she could do to improve her situation.

Now it's your turn to continue building this same awareness for yourself.

EXERCISE: Understand Your Body's Stress Response

Pause and reflect. Are you stuck in a vicious cycle where you can't achieve what you want to achieve and where your response, numbing out and criticizing yourself, feeds the beast by increasing your body's stress response? For the next week, track the following:

- How many hours do you sleep each night?

- How often do you forget something?

- How is your nutrition? Track when you're eating a balanced diet of vegetables, fruit, protein, and whole grains, and when you're not. Track when you numb out through overeating. Track when you skip meals and don't give your brain enough calories to function well.

- What are your energy levels at different times of the day? In the morning—high, low, or medium? Midday—high, low, or medium? Evening—high, low, or medium?

- How often do you space out or daydream? Hourly, daily, or not that often?

- What is your level of distractibility? When you receive new information, are you encoding it in your memory or entering it in your calendar, or are you immediately distracted by something else? How often does your mind wander, or do you daydream or think of things other than what's happening in the present moment?

- How much exercise are you getting? Do you walk or break up your day with any kind of exercise? Are you sitting a great deal? Are you always on your feet, never stopping to relax?

- What are your physical symptoms of stress? Heart beating fast, thoughts racing, body tension, tight chest, stomach pain?

Write daily in your journal, keeping track of your response to stress in these different areas. Jot down anything that stands out for you. Spotting where stress most impacts your life will help you focus on overcoming brain fog.

Now that you're more aware of your day-to-day experience of stress and brain fog, consider what you could gain by reducing the amount of stress you experience and overcoming your brain fog! Something is motivating you to make a change. After all, you picked up this book and are reading it. If you attain greater mental clarity, if you get back into your life, so to speak, what's in it for you if you achieve a more productive way of functioning?

EXERCISE: Reflect on Your Goals for Overcoming Brain Fog

Below are some common goals for overcoming brain fog. Check off any that apply to you.

- Greater work productivity

- Increased success at work

- Better grades

- Feeling more satisfied with life

- Becoming a better parent

- Becoming a better romantic partner

- Increased self-worth

- Increased control over your choices

- Increased peace of mind

- Feeling more present with your children, friends, and/or family members

- Feeling competent and effective as a person

- Feeling you can be yourself and authentic with others

- No longer feeling like a fraud or imposter

- More joy and laughter in your life

Feel free to write in your journal about certain goals that are especially important to you. If you have other goals not listed here, you may want to write about them as well, to further develop your awareness of what you hope to achieve by overcoming brain fog.

The awareness you're building now is the first step toward taking better care of yourself, dealing with the stresses that are currently driving you and your life, and breaking out of brain fog.

Putting It All Together

Chronic survival mode makes it close to impossible to pause and reflect. If you're stuck in brain fog, it's time for a psychological reset. This reset begins with self-reflection. Awareness, like the kind you're building now, begins a process that short-circuits survival mode. So, keep tuning into the impact of brain fog on your brain, body, and quality of life—and let's continue your journey out of this debilitating cycle. The next step is to look at ways to build a sense of competence and mastery where, right now, you might feel only helplessness.

SOLUTION 2

Overcome Helplessness with Mastery

You kill yourself trying to achieve the unachievable: to attend to every responsibility on your plate, to be the absolute best at your job, to be a perfect partner or parent, and to do it all looking and feeling good. Then when you fail, you beat yourself up and feel, as some of my clients put it, that you "suck at everything" or are "running in place" or "beating my head against the wall." Often, it feels like you're helpless in the face of all you're dealing with.

When our expectations for ourselves exceed our coping resources, we are made helpless. We are ineffective and we know it, and yet we can't bring ourselves to change the dynamics of our situation. There is too much inbound. If time would just stop, perhaps we could recover and develop a new strategy for going forward. The truth is, time won't stop, but we can take a time-out to assess how to better cope with and manage our life.

If you're in this dilemma, chances are you're reluctant or even entirely opposed to taking a break to reflect. Usually this conflict is not only about time but also about how hard it seems to accept that the way you're managing isn't really working. You might think that slowing down and making changes to the way you do things would be an admission of failure. And it's especially hard to admit this kind of failure when you feel that it's only by the skin of your teeth that you're holding yourself together.

So instead of pausing and developing new strategies for control and productivity, you keep going through the motions. You follow the same pattern day after day, even as you sense you're drowning. In the long term, this manner of existence is self-defeating because your situation never improves. But in the moment it's often preferable because it means you don't have to admit the reality that your life has probably overpowered you.

Noah's Story

When Noah came to therapy, it wasn't on his own terms. His wife told him to talk with someone or else she would leave him. Noah described a tumultuous past year. Before the pandemic, he had worked in an office surrounded by friends and colleagues. He traveled monthly for work and enjoyed the stimulation and novelty this brought to his life. He also loved his wife and two girls and always looked forward to getting back to them. He was often rewarded for doing a good job at work through bonuses, promotions, and accolades. And at home, his wife and daughters adored him, since he was an involved and caring dad. His wife was very happy in their marriage, which was a source of pride for Noah.

Once COVID hit, however, all of this was over. Noah and his wife both began to work full-time from their basement. The kids did online learning at home. Noah described his days as controlling him. He was typically going back and forth between work and the kids' needs, and it was a mess. He would respond to work calls while simultaneously trying to log his kids onto their remote learning platforms. Once he got back to his desk, he often forgot what he was planning to do when he'd left to attend to the kids. He would regroup, get back to work, only to be interrupted by his wife saying that she had a work call and that one of their kids was upset about missing an online assignment. He would drop everything to help, then return to work again scattered about what his goal was for the day. He often found himself surfing the internet or scrolling endlessly through social media to distract himself from the stress of trying to keep up—and the stress of living in the COVID era with no certainty about when it would end. By the end of the day, he would be disappointed with himself and feel entirely defeated. What's the point of trying? he would think to himself.

What's more, the pandemic not only increased the amount of work Noah had to do in his day-to-day life but also limited his options for taking a break from that work to get some relief from it. Noah no longer saw friends, exercised, or watched sports on TV—activities he'd always enjoyed in the past. He no longer had his usual coping resources for clearing his mind and keeping him focused. As a result, he often ended his days by overdrinking and passing out. Over time, this pattern led to more and more exhaustion for Noah and conflict with his wife. This further caused Noah to lose track of his focus and goals. He felt that he was failing at everything and believed himself powerless to make any kind of change.

When stuck in a state of prolonged stress, you too may end up falling into a routine of unsuccessful, unhelpful attempts at coping with that stress or persevering in spite of it. Repeatedly reenacting the same troubling patterns of behavior creates a self-fulfilling prophecy. You might come to believe that you truly cannot impact your future—a form of thinking that seals your fate to be nothing more than a cog in a wheel that goes nowhere. This sense of abject powerlessness is what psychologists refer to as *learned helplessness*. Learned helplessness means you no longer feel you can be successful in reaching your goals, getting your needs met, or enjoying and loving your life.

Learned Helplessness and Brain Fog

The symptoms of brain fog—feeling defeated and exhausted, with a reduced ability to focus and concentrate—set a stage for learned helplessness. We start to ask ourselves, *What's the point in trying if I am going to fail anyway?* Once learned helplessness sets in, people tend to give up. We stop trying to work within or master our circumstances, and we feel we can no longer achieve anything. In this state, we might feel entirely at a loss about whatever life is throwing our way—job insecurity or loss, loneliness or relationship instability, political uncertainty, difficult health issues—with no ability to fix it or the brain fog we feel.

In personality psychology, *locus of control* defines how much a person believes that they are ultimately responsible for their successes and failures. The word "locus" is Latin for location, and your particular locus of control comes down to whether you feel your control over your own life is internal, resting within yourself, or external, resting with forces you can't influence.

Locus of control has been extensively researched over the years. Having an internal locus of control has been shown to be a significant factor in whether we adopt pro-health behaviors, experience emotional stability, feel satisfaction in our relationships, and attain professional accomplishment. If you have a high degree of external locus of control, like Noah does—if you feel like your life rests mostly in other people's hands, not your own—you might be more prone to brain fog as well as other symptoms, like depression and overdrinking, that can arise from attempts to cope with your feelings. Adjusting your locus of control, learning to take responsibility for the things you *can* control, including the ways you deal with the circumstances that cause you stress, can help you deal with your brain fog in ways that are masterful—not helpless, resigned, or self-punishing.

It's important to note that having a high level of internal locus of control doesn't necessarily mean that you'll have perfect control over your own life. We all inevitably deal with moments that are so much bigger than we are that we lose some of our ability to make our own decisions or control what might happen to us. Likewise, having a high external locus of control doesn't mean you are a worse or weaker person because of it.

Many factors impact whether or not we develop a high internal locus of control. How your family or caregivers reacted to your successes and failures as you were growing up is a significant factor. For example, praising your successes without emphasizing or rewarding effort and hard work might lead you to develop a high external locus of control: your sense of self-worth rests on the judgments of others and whether those are positive rather than on the efforts you make. Also, if your caregivers blamed external events for their own life course, or didn't take enough internal responsibility for their decisions and actions, they may have taught you to hold a higher external locus of control. And, if you genuinely struggled with events outside of your control for periods of your childhood and adulthood, including socioeconomic status, trauma, abuse, war, or social unrest, you may be prone to having a high external locus of control.

If you struggle with brain fog, it may be that you have a high external locus of control, or maybe you've actually experienced and benefited from a healthy sense of control and agency for most of your life. Often, when we are hit by one too many stressors, we lose our balance and no longer feel as if we have the control we once had; the locus shifts from internal to external, and then we give up.

One way to start understanding and dealing with your brain fog is to become aware of whether you're attributing your state of mind to factors outside of your control. For example, do you attribute control of your successes and failures to yourself or to a fated force? If you continually find yourself feeling angry, resentful, or upset by the events in your life, it would be helpful to reflect on whom you blame for your ups and downs and what that might say about how capable you feel yourself to be and how much responsibility you take over your own life or actions. How you internalize a particular point of view about control speaks volumes about your ability to feel productive, focused, motivated, and present.

EXERCISE: Reflect on Your Locus of Control

Who is in control of your life? Respond yes or no to each of the following questions.

1. Do you believe positive events in your life are mainly due to luck or chance?

2. When you hit a setback or fail at something, do you blame others?

3. When you are upset, do you feel as if your emotions are out of your control?

4. When you have an argument with a friend/romantic partner/family member, do you repeatedly tell yourself what they did wrong?

5. When you hit a roadblock or challenge (interpersonally or professionally), do you tend to give up, that is, want to break up or switch job assignments?

6. Do you feel you're just going through the motions of whatever your day throws your way?

7. Do you blame events (politics, global events, work stress, family adjustments, divorce) for your state of mind?

8. Do you repeatedly engage in the same behaviors, day in and day out, even though they keep you stuck?

9. Instead of holding yourself accountable, do you make excuses for your state of mind? These might be "I am stressed out," "I can't do anymore," "I am too tired," "I am too busy," or "People expect too much of me."

10. Do you believe that hard work, getting out of your comfort zone, and adopting new coping strategies, are not worth the effort?

Answering yes to all of these questions suggests you have a high external locus of control; answering yes to a few suggests you externalize control in some situations. If you're externalizing more than is healthy or useful to you, try to find ways to change to a more internalized locus of control.

When he was completing this exercise, Noah could clearly see he was just going through the motions, dealing with roadblocks and feeling brittle rather than flexible and resilient. He was also blaming COVID for his state of mind rather than realistically acknowledging that, yes, the pandemic had made life hard and challenging *and* there were still things he could do to deal with the stress it was causing him, things he just wasn't doing at the moment.

Consider if you've made excuses for your circumstances and told yourself that your way of functioning is a product of forces outside of your purview. If so, consider that perhaps, in the fog, you've overlooked a different path—a path that is available to you and one in which you can feel competent, be energized, and obtain the larger goals you have for yourself. The more in control you feel, the more grounded you become. If you put effort into growing and coping with life in more adaptive ways, it will pay off.

The Self-Pity Trap

Of course, whether we are or feel in control of our lives isn't the only problem. When our circumstances become more than we can health-fully bear, at least with the ways we're currently responding to those circumstances, it's naturally tempting to turn toward self-pity. When you're in a tough, stressful phase of life, and you feel like there's no hope for improvement, you might give up and feel victimized by life. This is an understandable response, but it only exacerbates the diffi-culties at hand, compounds the stress you feel, and makes it that much harder to liberate yourself from brain fog.

You're not alone in dealing with tough times and situations that are hard to control perfectly. Life often involves dealing with things as they come, outcomes we can't always predict or control for, and people who have their own minds and needs. Certain cultural, political, or economic events can also have tremendous power over our lives, power we can sometimes struggle to deal with. Chronic stress and brain fog are understandable responses to these moments. Feeling bad about all of this is also understandable, especially when you've been in brain fog for a while, but it's important to resist self-pity. The solution is to use the internal locus of control you're beginning to harness, focus on which parts of your situation are within your control, and see what you can do to change them.

EXERCISE: Recognize What Can Your Control

Here is a list of some things that may be more in your control than you've recognized. Consider each item on the list and whether you could do more to improve the way you behave and respond to stress in that area:

Your reactions

Your health behaviors

Your peace of mind

Your self-worth

Your attitude about change

Your nutrition

Your physical exercise

Your free time

Your boundaries with others

Your self-care

Your negative thoughts

Your anxiety

Your pleasure

Your mood

Your sense of joy

Your ability to love

Even in the most difficult, overwhelming situations, there are things you can do to deal with what you're facing, aspects of your situation or how you're interpreting and responding to that situation that you *can* control. Can you think of other ways in which you may not be as helpless as you feel or have told yourself?

Throughout this book we will talk about actions you can take to become more in control of how you interpret and respond to events or situations. We're focusing on behavioral change for a reason. As neuroscience reveals, any behavior you do fires certain neurons in your brain. For behaviors you do regularly, the same neurons fire regularly too. And if neurons fire regularly often enough, the behaviors in question become wired in your brain. This is why certain ways of responding to your situation that may not be helpful in breaking you free of brain fog—like Noah's surfing the internet to distract himself when his feeling of stress is intense—nevertheless have become intuitive for you: an automatic, instinctive habit. It also means that when you change your behavior, it actually changes your brain. When you behave differently in a consistent way, your brain changes, and that change actually makes the new behavior easier and more instinctive.

How Mastery Impacts the Brain

Donald Hebb, a prominent psychologist in the field of neuropsychology, worked extensively to understand the role of brain neurons and learning. It was Hebb who developed the theory that the brain changes as a result of new experiences: when new learning occurs repeatedly, the neurons involved develop a stronger connection and will fire more easily in the future. This is how new learning experiences are remembered: they are sewn into the fabric of the mind over time. It's also why your brain, seeking to expend the least energy possible, repeatedly triggers the old, familiar, and thus easy circuitry that already exists. Thus, although it's hard on you to remain stuck in brain fog, it's easy on your brain; when faced with stress, the brain just calls upon those already existing, well-worn neuronal patterns.

The encouraging news is that neuroscience shows that our brains are much more flexible than perhaps we realize. In fact, when presented with repeated experiences over time, our brains literally adapt and grow accordingly. Athletes, musicians, and monks have all increased the capacity of their brains for their specialized pursuits: playing a sport, playing an instrument, and meditating. Individuals who lose hearing or vision develop increased neuronal connections for their other senses; for example, people who are deaf often have an increased capacity to attend to what's in their peripheral vision (Bavelier, Dye, and Hauser 2006).

Neuronal plasticity means that changing your behavioral habits and thinking patterns changes the brain's wiring on a physical level.

This also means that changing a behavior that's been set can be difficult, at least initially. If you've always taken the same route to the grocery store, it will take conscious effort to train your brain toward a new route. It can be done, but you have to be deliberate because your brain would prefer to just keep it easy and stick with that old familiar way. But once you take the new route for a while, and new neurons have been given the chance to fire often, it becomes effortless.

If you're stuck in brain fog right now and in a state of learned helplessness, you can take comfort in the fact that ultimately, you *can* rewire your brain. The synaptic pathways involved in attention, thinking, motivation, and concentration can all be altered by new learning over time. And the brain changes when it's made to work repeatedly at something new. The process by which one develops brain fog—your brain responds to experiences in particular ways, which over time harden into routine—is the same process that will help you to become focused and calm once again. Positive change will come as you repeatedly engage in new habits of coping, through changing your thoughts and behaviors.

Self-pity comes when you're feeling helpless, out of control, with no way out of your brain fog, sensing that *things will never change.* To begin setting yourself on a path to new wiring, new behavior, change, and agency and ease in your life, let's start by imagining what your life would look like if brain fog were no longer a problem for you. What would you be doing if you weren't stuck and helpless like you currently are, but instead exhibited mastery over helplessness?

EXERCISE: What Life Would Look Like Without Brain Fog

Brain fog is so distracting and exhausting that you may feel you don't have the mental strength to even consider your larger goals. Still, see if you can take a moment now to imagine that the constant stress and feeling that you're just treading water is over. You are on good terms with your family and loved ones. The circumstances in your personal and professional life have been resolved. You no longer feel like a fraud and powerless to make things better. The weight of all that has been lifted. You're in control. What would your life look like? How would you feel? And most importantly, what would you be doing?

Take a moment to write in your journal about this, using the following prompts if they're helpful.

If I didn't feel so stressed out all of the time, then I would be doing...

If I didn't feel so scattered, then I would start...

If I felt capable and organized, I would...

If I could focus, then I would...

If my mind was at peace, I would start...

Based on the life you envision for yourself if you weren't managing chronic stress, set a goal for yourself. Write it down in your journal.

You can begin the process of rewiring your brain just by setting this goal for yourself. You can start with a big or a small goal. What matters is that you stick to your goal even when life stress and brain fog threaten to set back in and throw you back into helplessness.

As an example, when Noah considered life beyond brain fog, he eventually realized that if he was going to continue working from home and have fewer perks of being in the office and traveling, he would probably enjoy starting his own company. Initially, Noah wouldn't even allow himself to go much further with that thought. He was entirely overwhelmed just by the idea. "I can't even imagine taking on something like that right now," he said. However, the thought did light a small spark. Noah agreed to start spending fifteen minutes each day, alone, quietly breathing and reflecting on what was going on in his body and his mind. He committed to this practice and did it most days. This small space for himself and his needs started a larger process of growth for Noah. He picked back up some of his old activities, including exercising, began communicating with his wife about hiring a caregiver for their children, and eventually started to feel more capable to take on the larger goal of starting his own company.

You can retrain your brain through focused exercises like the one Noah did.

Retraining Your Brain

The brain has many ways to defeat our best intentions. You know what you need to do, and yet you can't muster the mojo to just do it. Maybe you sit down to start thinking through your life in another way and, before you know it, you're scrolling through Instagram or looking at cute pictures of puppies on the internet. The day ends, you're defeated, and you wake up in the same rut the next day. The reality is you tend to procrastinate because you're not in control of your mind, but you can learn to focus.

EXERCISE: Fifteen-Minute Daily Focus

Consider committing to this daily exercise to begin training your brain to focus, not just on checking boxes but on the deeper goals you have for yourself. It's quick and easy—but you'll need to do it every day to start. Set an appointment in your calendar, program a reminder into your phone— whatever it takes. Literally force yourself to practice this fifteen-minute focus once a day, and do it consistently.

1. Set a timer for fifteen minutes.

2. Review your writing from the previous exercise, in which you imagined what life would look like beyond brain fog, and set a goal for yourself.

3. Reflect on the baby steps needed to achieve the goal of what you would be doing if you weren't so chronically stressed out.

4. Take one actionable step toward reaching your goal. A step can be something you focus on during this fifteen minutes— for example, working on an Excel spreadsheet to log your bills or finances, doing a job search, researching a hobby you'd like to start, calling your accountant, or setting up a website. It can be taking a brisk walk, calling an old friend and reconnecting, spending time with your partner or child with no agenda but just being with them, or spending this focused time breathing. A step can also be setting up a structure or time in your calendar to do tasks that require additional time or will be done on a regular basis—such as walking every Tuesday and Thursday mornings, emailing and setting up a time to have lunch with colleagues to brainstorm job opportunities, taking an academic class, or participating in a monthly book club.

5. Make a commitment to yourself that you will achieve at least one actionable step each week toward your larger goal.

6. At the end of the fifteen minutes, move on to the rest of your life.

Come back to this practice each day so that you don't lose sight of what you really want out of your life.

As you make space for this kind of reflection, you will find that your goals will evolve and sometimes change. You may come up with new goals. That's okay. What's important is that you are making room in your life to consider your deeper yearnings and to take actionable steps.

Putting It All Together

It may not always feel this way, but in the end, you choose which way you wish to control your life. You can keep engaging the *Groundhog Day* existence that brain fog, with its learned helplessness and self-pity, so easily puts us in—and your brain's wiring will remain the same. You will stay stuck in an unfocused and frazzled state of mind. But if you allow it to, the brain has an incredible ability to adapt and grow. The key is to just give it new experiences. Starting with fifteen minutes of focused time every day, give your brain opportunities to start rewiring. Like beginning a new fitness regime, doing this is difficult initially—but with time, the routine becomes easier and easier. Next, let's look at ways to build connections to others so you can overcome the social isolation and loneliness that so often characterize brain fog.

SOLUTION 3

Overcome Social Isolation with Attachment

If you struggle with brain fog, there's likely no true witness to the hardship you face, no one who really sees you and knows the depths of what you're managing. You may be around people and act the part of a social person, but in truth you let people in only on a surface level. Perhaps you fear the judgment of others, if they were to know the real you, or you don't feel you have the time or mental resources to deal with people, or deep down you don't believe anyone would ever really be there for you anyway, even if you tried to let them in.

Brain fog will trick you into believing that investing in your social world will merely invite more turmoil. Like a passenger on a boat on the verge of sinking, you jettison excess weight to stay afloat for as long as possible. And relationships may feel like too much weight to carry when you're on the verge of sinking. What gets overlooked in this panic is the emotional consequences of not having close others in your life. Instead of helping you reach a safe shore, lack of connection keeps you cast out to sea, adrift.

In fact, when your boat is sinking, not investing in your relationships is similar to throwing a life jacket overboard. Relationships are as vital to our survival and success as shelter, food, and water and have the power to ease the worst of circumstances. Civilizations would not have survived, indeed humankind may not have survived, if we did not depend on, connect with, and cooperate with others.

As this chapter will explore, our brains are wired for connection with others. When this natural human need for connection goes unmet, we no longer experience the rewards of personal relationships. As a result, we slide toward the profound agony of loneliness. We become irritable and depressed. And, without this closeness, we have no one with whom we can brainstorm, share our perspective, and gain new insights, so we live unfocused, overwhelmed, and confused about what steps to take to mitigate our circumstances.

Social disconnect does not happen all at once. As life throws you another and another stressful curveball, you inch further and further away from close interactions. And as meaningful interactions decrease, negative and self-critical thinking increases.

Cassandra's Story

Cassandra was a fixer. She could make anything in her life better: her husband, her job, or her children. When a problem came up, she'd go into Ms. Fix-It mode. Her husband knew this about her, and in Cassandra's mind he had benefited from it for years. Then out of nowhere, her husband of fifteen years told Cassandra that he was no longer in love with her. Cassandra was thunderstruck. She avoided her feelings and instead went immediately into fix-it mode.

Cassandra spent the next six months trying to convince her husband that things between them could get better. She tried to dress differently, be sexier, funnier, complain less. They even went on a couples therapy retreat. All to no avail. When Cassandra's husband finally left, she stood in their empty living room in complete disbelief. She couldn't believe this was really happening to her.

Initially, Cassandra was so overwhelmed that even she couldn't stop herself from crying in front of her parents and close friends, but after the initial shock wore off, she stopped opening up. As a result, Cassandra was alone with the inordinate burdens of working full-time, raising two young children, and seeing her way through the legal logistics of divorce and custody arrangements. She had no one to support her or with whom she could confide or talk through the myriad of conflicting emotions she faced.

Over time, Cassandra increasingly felt burned-out. Her brain was constantly in overdrive—thinking, planning, scheduling, second-guessing herself—at the same time she was stuck, unable to effectively problem solve and make decisions. After a few minor fender benders and finding herself one day brushing her teeth with hydrocortisone cream instead of toothpaste, Cassandra became concerned she had a neurological condition. She went to her primary care doctor and shared that she couldn't think straight and that there was something wrong with her brain. After undergoing a series of medical tests, Cassandra was pronounced healthy and referred to me for stress management.

In our initial meeting, Cassandra ashamedly declared that she simply couldn't focus. In fact, as we spoke, her thoughts often trailed off in various directions or she couldn't remember her point or what she was trying to say. It was easy to see that Cassandra was managing way too much and without any real support. Her lifelong commitment to uncompromising self-reliance was holding her back.

Cassandra had a lot of people in her life—friends, parents, a close cousin—why not turn to them for real support? It wasn't *that she hadn't considered doing so; she thought all the time about opening up, and a part of her knew it would be a relief to share. But each time she considered confiding in a specific person, she hit a block: with her parents, it was* I don't want to worry them; *with her oldest and dearest friends,* I was the one who had it all together. They'll think I'm a failure; *with her fellow mom friends,* I don't want them to treat my kids differently as a result.

Cassandra certainly saw people. In fact, as an elementary school teacher, she was around kids and other families quite a bit. In an attempt to seem "normal" and not be a burden to others, she stuck with sharing only the basics of her life or made funny jokes about her situation. For example, when talking on the phone, a friend might hear her kids screaming in the background, and Cassandra would say, "Living the dream over here." Friends and family thought she was okay. But she was anything but okay.

In fact, Cassandra feels extremely abnormal and like she doesn't belong anywhere. When not in the service of her children or her job, she finds herself ruminating about all of her own flaws. She feels like a failure. Her thoughts race with what to do, how to handle life, and how to feel safe again. She feels helpless and alone. At the same time, Cassandra's commitment to self-reliance has told her that it's weak to need other people and that she should be able to sort out her situation on her own. Hiding her emotions from others while feeling entirely overwhelmed has created internal tension for Cassandra that is blocking her ability to be present in her life—with her kids, with her family, with her work, with her friends, and most of all with herself.

Cassandra is experiencing detachment, a common symptom of brain fog.

Living with Detachment

Detachment is on a continuum and can range from subtle disconnect to full-on isolation. As an example, you may be around people in body but not in mind, as you're not truly connecting or being present. Or you listen to others, but you don't share your hardships and vulnerabilities. The reverse can also be true: you talk nonstop, so others in your life don't feel heard. As another example, you may have a family but you go down to the basement to watch television every night after dinner, never connecting with your children and avoiding your spouse. Or you may overschedule every aspect of your life, never really pausing to rest in the moment. More extreme examples of detachment include never leaving your home because everything feels too overwhelming or having so much social anxiety that you never allow yourself to make a new friend or start a new pursuit.

Wherever you are on the detachment continuum, if your basic needs for human interaction are going unmet, then you're likely turning to other, self-defeating outlets for a sense of connection. For example, the euphoric effects of drugs and alcohol often bring along an immediate sense of belonging. Of course, these effects wear off quickly and often leave people feeling more alone than ever. Similarly, materialism or buying things brings along a hit of joy that is also short-lived and leaves people needing to buy more to get another hit.

Consider for a moment the subtle or not so subtle ways you disconnect and whether you turn to other quick fixes over working at close relationships, romantic partnership, and building a sense of community. Here are a few more examples of the fool's gold that can make us feel momentarily connected but in the long run feed social isolation and disconnect: compulsive social media use, alcohol, drugs, compulsive shopping, excessive binge-watching TV, obsessive focus on appearance or unnecessary plastic surgery, judging others, gossiping as a way to connect, excessive control (emotional or physical) over others, only sharing your good experiences and strengths with others, only talking about yourself, overly attending to others' issues but not your own, workaholism, and compulsive preoccupation with business interests.

These behaviors may make us feel better in the moment, but in the long term exacerbate not only social isolation but also cognitive decline.

Why We Need Relationships

Whether you have close relationships in your life or not, chances are you think about people a lot. You consider what others want from you and what you ultimately want from them. You imagine judgments that people make about you and perhaps make your own judgments about them. You may beat yourself up for what you perceive as your flaws in your relationships or imagine others would not like you if you showed them your real self.

According to social neuroscience expert Matthew Lieberman (2014), when our brains are not engaging an active task and are at rest, we default to "mentalizing" or playing out various relationship scenarios, including the thoughts, feelings, and motives of the people in our lives. Basically, when the brain has a free moment, it tries to make sense of ourselves and our social world. This tendency to think to this extent about our social selves is nothing short of astonishing. Mentalizing is evolution's way of ensuring that we don't lose sight of how important other people are to our survival.

The often-referenced psychologist Mary Ainsworth developed foundational research that shows how babies begin to attach to their primary caregiver at birth. It's a basic human drive from the beginning. Moments after birth, newborns grasp their caregiver's finger, and they sleep more easily when pressed against the warmth of another. Needs are eased with parental responsiveness, and if attachment to caregivers goes well, brain growth flourishes.

Even if you're living alone, very independent, or introverted, the need for social connection to maintain mental health is still present and endures through adulthood. Having close others in our life helps us to maintain a realistic perspective and stay optimistic and helps to regulate our emotional experiences. When our emotions and needs are reflected in the faces of our partners, friends, neighbors, family, and colleagues, we feel safe and grounded. When we feel heard, that we belong, that we're needed, that we're seen, and that we're valued, our nervous systems let out a deep exhale. This is because when our relationships are positive, they actually interrupt our negative thought spirals and decrease our body's stress response.

When our relationships are unfulfilling, or we can't feel safe in close relationships, or we feel like there's nowhere we belong, we crave this missing and essential component of our survival. We replay negative thought-streams regarding others and ourselves as an attempt to make sense of or even justify our situation. All of this brings on a feeling of defeat and a tendency to see more of the bad than the good in the world.

As you read through the next couple of sections, take a moment to reflect on the content of your thoughts as they relate to other people. Do you fall into one of these common fallacies of detachment thinking?

Believing You're Better Off on Your Own

Telling yourself that you can, and should, manage your hardships on your own is enticing because our culture shames helplessness and conflates an independent can-do attitude with strength and resiliency. Handling it all on your own may also bring comfort and pride and protect you from the vulnerability that comes when you open up our private, intimate worlds to others. Because of these immediate rewards, your rationalizations work and you remain stuck in your bubble. In the long run, however, you've cut yourself off from the one thing that can truly heal—close, meaningful relationships with others.

Here are some common examples of rationalizing social isolation:

People are drama.

I actually hate people.

Ugh, people are too much work.

I am happier on my own.

No one can make my situation any better.

I need all my energy to deal with my crazy life, and people are too needy.

I am better off alone.

It's an absolute myth that we can healthfully get through life on our own without meaningful connection. It's only by being vulnerable and letting others in that they can come to know and feel close to us and we can become stronger.

Fun fact: research shows that when we feel helpful to others, the reward system in our brains light up. So let others in; they need it as much as you do.

Believing You Don't Fit In

Not maintaining regular, close, social interactions means that your social muscles are weakening. You'll have a harder time picking up the subtleties, nonverbal signals, and intuitive or instinctive pieces of social interactions. And not seeing people regularly—or not being present when you do see people—brings on a sense of not belonging and a feeling that people don't care or want you around. When this happens, you're more likely to misinterpret the motives of others or see others as being against you. Instead of recognizing that interacting socially is a skill base that can be put back together, you second-guess and criticize yourself. This self-defeating loop makes you feel bad about yourself, so you want to avoid relationships all the more.

Here are some common examples of blaming yourself or others that bring on anxiety and fear and can block you from being your full self with other people:

I don't fit in.

Why did she look at me like that?

What did he mean when he said I look different?

She was glaring at me the whole time.

Oh no, that went horribly.

They must think I'm an idiot.

I can't believe I said that. What do they think of me now?

Why am I always on my own, and others have people to support them?

For most people, these negative thoughts decrease the more they practice connecting. Literally showing up, being present and real, and doing it over and over again breaks the pattern. Your social skills will improve, and you'll become more comfortable sharing and being open. The amazing thing about human connection is that it only takes one meaningful experience to bring on some momentum.

How Social Connection Impacts the Brain

Our ongoing need for connection means we talk, date, exchange long texts, support our friends and family, scroll social media, watch concerts and plays, enjoy participating in sports teams and clubs, help people in need, turn to others when in need ourselves, parent, volunteer, engage in small talk, love our pets...the list is long. When the drive to connect is met, the stage is set for increased pleasure and mental sharpness.

In an effort to better understand health and happiness, researchers from the Harvard Study of Adult Development (2015) collected reams of data (including medical records, questionnaires, and interviews) on a group of Harvard students, their offspring, and a group of inner-city Boston youth. Tracking the original cohort, often referred to as the Harvard Grant Study, began in 1938 and consisted of 268 Harvard students. The pivotal outcome from this influential study is that close relationships keep people happy throughout their lifespan, help delay mental and physical decline, and are better predictors of long and healthy lives than social class, IQ, or even genes.

Interestingly, the Harvard study demonstrated that those participants most content in their relationships at age fifty were the same participants who were physically healthier at age eighty. In addition, those who in their eighties were satisfied with their marriages reported that even when they experienced physical pain, their mood was unaffected. Folks who did not report feeling happy in their marriages, on the other hand, felt greater physical and emotional pain. Research shows that when we have supportive listeners in our life, we're more likely to have increased cognitive functioning, mental sharpness, and brain health (Salinas et al. 2021).

These findings are consistent with the all-too-common pattern of an elderly person going into cognitive decline after losing their life partner. Perhaps one reason for the decline is that loneliness is as painful as physical pain and, as such, is so distracting that we cease to exist in the present but rather become stuck in our heads. In fact, when research participants are exposed to social pain, the part of their brains that registers physical pain becomes activated (Lieberman 2014). And most fascinatingly, when participants are given Tylenol and exposed to that same social pain, the brain regions that register physical and social pain no longer show this elevated response (Dewall et al. 2010). In effect, physical and social pain both hurt and can both be all-consuming.

If you accidentally slice your thumb while chopping vegetables, all you can think about is getting the pain to stop. Similarly, if you're isolated and lonely, all you can focus on is your misery and wishing for relief. The sliced thumb heals quickly. Sadly, without some kind of intervention, loneliness is endless. Being fixated on the pain makes it almost impossible to be present, to focus, attend, and remember...let alone be happy.

Our brains are wired to trust that we'll have people around us in whom we can confide and on whom we can rely. When this core assumption goes unmet, our brains are at a loss. In the absence of positive social interactions and close connections, we're aware that we're out of sync and then become preoccupied by the reasons, causes, and possible cures. On the other hand, when experiencing rewarding social interactions—whether giving to others or receiving support ourselves—we feel once again safe and grounded.

Cassandra Rejoins the Herd

For Cassandra, healthfully adjusting to divorce meant rethinking her lifelong commitment to managing difficult emotions through a self-sufficient, can-do will-do attitude. She now realizes that she was uncomfortable with vulnerability and tended to judge others as "needy" and "weak" when they communicated their difficulties. A large part of Cassandra's identity was a belief that she was the one who had it all together. She once felt pride in solving her problems all on her own, but that pride also blocked emotional connection. Over time, Cassandra began to assume she was the only one with problems. In reality, because Cassandra never expressed vulnerability, the others in her life felt they couldn't let their guard down either.

Even Cassandra couldn't will herself out of her state of distress and upset. This brought on self-critical thinking (What's wrong with me that I can't get it together?) *and panic* (If others see the real me, they won't like me anymore). *Cassandra shared with me that she literally felt like an "alien passing for human." She'd see people and play the part of normal but inside felt cast out and directionless.*

Cassandra's experience in therapy was a start to a new pattern. In opening up, she didn't feel weak but actually felt lighter and more grounded. She also began accepting that most everyone experiences difficulties in life. She wasn't alone. Gradually, Cassandra started to confide in a fellow teacher who had gone through a divorce ten years earlier. This one connection helped her to no longer feel like an alien. Through their talks, Cassandra became more secure with what needed her focus and what she could let go.

On the morning her divorce was finalized, Cassandra found her fellow teacher and friend waiting for her in her classroom with cake and coffee. As they toasted their coffees together, Cassandra's friend looked her in the eyes and said, "You did it. The worst is behind you!" Cassandra teared up, not because she was divorced but because her friend understood what she had been through and loved her all the more for it.

If brain fog has been getting in the way of your relationships, you too can set a goal for yourself to cultivate deeper connections with others. You need to make room in your life to invest more fully in relationships. To be successful, you'll need to be intentional and deliberate. Prioritize relationships and connecting by making a point to put it in your schedule every day. If old mental habits tell you that *Relationships are a waste of time,* remind yourself that your longevity and cognitive functioning are a stake.

EXERCISE: Cultivate Deeper Relationships

Here are a few ways to start expanding your relationship world. Find ones that appeal to you and put them into action:

Call an old friend and be honest about how you're doing. What do you have to lose?

Commit to a weekly volunteer pursuit and show up each week, even if you don't feel like it.

Join a group. Groups can instantly make you feel as if you belong somewhere. It could be a book club or a gardening club, a language class, volunteering for the PTA, attending church or synagogue or a meditation group, or joining a running group, a hiking group, an antique collectors' group, a dog-walking group...the list goes on.

Try to see one or two people at least once a week. You have no agenda except to listen and be open.

Take fifteen minutes to stop and chat with your neighbor.

Say yes to social outings.

Make small talk. It's not a waste of time; small talk is how deeper connections begin!

Take a walk with a friend and talk about your brain fog and what you're working on to get out of it. (You'd be surprised by how many people can relate.)

Note that when connecting, it's not enough to just show up and distractedly look at your phone or daydream. Work to bring your full self to the table. Focus on what others are saying, then listen and reflect back what you hear. Make a point of saying something of substance about yourself in every interaction.

Fun fact: you're very likely to receive a brain boost if you exercise while interacting with someone—even a coach, a trainer, or a stranger you pass as you're out walking. This is because exercising and socializing both trigger the brain's reward chemicals.

Putting It All Together

Listening and sharing with others is an extremely effective step in pulling out of brain fog. And the more you do it, the more you'll reap the rewards. At the same time, the simple act of talking requires you to start connecting with deeper parts of yourself. As a result, difficult emotions, which perhaps you've pushed away for a long time, may surface. Knowing what you're feeling on a deeper level will help you develop emotionally intimate relationships. Also, spending time understanding and coping with your emotions relieves the body's stress response. Let's turn now to increasing your emotional awareness.

SOLUTION 4

Overcome Survival Mode with Emotional Awareness

If you avoid your feelings regularly, it's definitely contributing to your brain fog. This is because the more you avoid your emotions, the harder it becomes to know how you feel about the various aspects of your life, and over time you may stop knowing who you are entirely. When we push away the negative emotions, we cease experiencing the good ones too. Joy and pleasure become elusive because we aren't able to connect with ourselves and our experience of much of anything. Brain fog creeps in when you no longer reap pleasures and rewards from life or feel connected to your experiences.

As the unprocessed emotional baggage mounts, you start to experience psychological symptoms—panic attacks, crying spells, obsessive thinking, angry outbursts, fear, and anxiety. Symptoms such as these make it impossible to focus, concentrate, and be present. As time goes on, you likely have little sense of where the original negative emotion started or why you're feeling so entirely awful and overwhelmed.

Although emotions are not actual things we can touch, they wield immense power over our lives. Avoiding your emotional world is akin to ignoring a check-engine light illuminated on your car dashboard. You see the light, but you tune it out and just keep on driving. Eventually you're on the side of the highway with smoke billowing out of your hood.

Duncan's Story

Duncan wasn't aware that he was avoiding his emotions when he came into therapy one day absolutely petrified. For three months, a racing heartbeat, cold sweat, shaky hands, and a stomachache would startle him awake at night. Each time this occurred, Duncan believed he was dying of a heart attack. As the pattern continued, he became more and more preoccupied by a fear of going to sleep. For hours, Duncan would walk around his home in a state of morbid drowsiness.

Lack of sleep only worsened his fears, and soon just the sight of his bed induced panic. He couldn't focus or concentrate on his work and felt entirely ungrounded—like a sleepwalker—during the day. Athletically fit, tall, and intelligent, Duncan had always felt stronger than his peers. This confidence had allowed him to typically push through difficult emotions or avoid them altogether. For the first time in his life, Duncan was coming undone.

The anxiety started to extend into the rest of his life. Duncan reported that any small thing in his life—a work project, a changing deadline, injuring his elbow, forgetting his phone—would send him into a tailspin. He'd immediately jump to the worst-case conclusion: I can't make this deadline. I'm going to get fired. Or he would think, I may need to get elbow surgery. Or, upon misplacing his phone, I can't do anything without my phone. This catastrophic thinking reinforced his fear spiral and kept him constantly on edge.

Duncan tried everything before coming to therapy— nutritionists, acupuncturists, physicians, yoga trainers, a shaman, a hyperbaric chamber, and medical marijuana. In reality, he was trying anything he could to continue to avoid his emotions. He believed that psychotherapy was for "crazy people" and people *who just want to "talk about their feelings all of the time." But things had gotten so out of control that Duncan felt he had no choice but to seek therapy. With fear in his eyes, Duncan pleaded with me to "please...please...help."*

Duncan revealed that his eleven-year-old daughter had been diagnosed with a chronic medical condition around the same time that he was forced to take on a new role at work. He had never processed the impact of these stressful life events, and, unfortunately, not processing his feelings only brought about more negative emotion, which eventually sent Duncan into a crisis state. This was not Duncan's intention. He had been doing his very best to stay afloat.

As Duncan story illustrates, brain fog increases when we allow ourselves to sidestep our feelings. For some people, the pattern is a result of stress: you just don't have the time and energy reserves to consider your feelings. For others, this emotional avoidance is a learned habit that may have started in childhood. Perhaps your caregivers didn't help you to understand yourself emotionally and how to cope with strong emotion, or they did not model healthy emotional coping. Whatever the reason, emotional avoidance is self-defeating because it only makes it harder and harder to focus, remember, and feel at peace.

Defeating Yourself with Avoidance

Avoiding your emotions makes it hard to focus, because it takes a lot of energy. You're on high alert 24/7, lest that dreaded emotion comes pouring in. You likely keep emotion at bay through some version of an overthinking–overworking spiral, where your brain is so full of distracting thoughts that there's no space to know your deeper feelings. You may overthink mundane issues, events, and details without any of them being the real problem. This is akin to focusing on whether it's going to rain when the check-engine light is flashing.

This is why panic attacks result. You relax for a moment, while you're sleeping or driving or doing some everyday activity, and poof, you're in a full-on emotional meltdown. Of course, panic just reinforces the sense that you can't relax, which adds further layers of stress.

The pattern of avoidance doesn't just manifest in intermittent panic attacks. When you avoid or push away your emotions, they always resurface and always in self-defeating ways: passive-aggressive behavior, compulsive working, obsessive thinking, compulsive eating, obsessive dieting, or drug and alcohol abuse. You may be prone to extreme emotional reactivity: meltdowns, crying spells, acting out in anger, and overthinking spirals. Alternatively, you may walk around completely flat, numbed out, seeming strangely calm and detached.

It simply feels easier, in the short term, to keep your thoughts on repeat, docking them on random places—kids' schedules, to-do projects, possible worst-case scenarios, health anxiety, worry about friends or family—than to actually pause and identify the root of your feelings. At the same time, when you don't let yourself recognize your deeper pain, you keep yourself in the dark about whatever the underlying emotion is that's at the source. So you walk around unhappy and distressed with no access to the cure for that distress.

In fact, the cure is to tune in with yourself. Tuning in with yourself will actually give you energy and reserves needed for other pursuits. Tuning in allows you to be present, remember, focus, and achieve.

There is no one right way to feel—feelings are what they are. If we felt nothing or neutral, life would hardly matter, and we grow as we give airtime in our minds to our more difficult emotions. Knowing and tolerating our painful feelings helps us to understand our preferences, goals, desires, and objectives; this process in turn brings about increased meaning and purpose.

If you're tuning out this critical data source, then you likely have enormous blind spots about yourself and what's happening in your life and with the people you love. Without accessing that data, you're walking around with a handicap that will make it hard for you to connect with others and almost impossible to feel content with yourself and your life.

Tuning in with Yourself

Consider for a moment whether you've bought into the myth that your feelings are unpredictable, scary, enduring, and have the ability to make you crazy. Like many, you may believe that if you acknowledge your emotional pain, you will become even more distressed. This is not your fault. Popular media, well-meaning friends and family, and even some therapists will give you strategies for avoiding your emotions so that you don't have to feel bad. The cure is actually quite the opposite: there are no shortcuts to knowing yourself emotionally; you have to feel your pain to no longer live in fear of it.

It's uncomfortable to experience negative emotion, but think about it: have you ever experienced a painful feeling that never left you or never became less intense? For the vast majority of us, the answer is no. This is because emotions are temporary—they come and go—and are much more predictable than most of us tend to realize. They follow a course. As you tune into a specific emotion and feel the upset in your body, it will become more intense but then subtly decrease in intensity.

Emotional intensity decreases as you give your feelings airtime in your head and meet them with kind, warm attention and curiosity. Your emotions are like a screaming child who screams louder when ignored but calms when given kind attention. As you become more skilled at noticing and not judging your feelings, you will find that your emotions no longer dominate and you have the mental space to focus and to enjoy your life. Here's an exercise to do just that.

EXERCISE: Emotional Tune-In

This is a road map for how to face your emotions without letting the discomfort overwhelm you, and it only takes ten minutes!

1. Give yourself ten minutes to imagine a part of you that observes, separate from your thoughts and emotions.

2. Be a curious observer. Notice where your mind roams and the accompanying physical sensations in your body. At first

you may think, *There's nothing. I feel nothing.* This is because we get used to avoiding the physical by going immediately to our thoughts. Yet there's a physical sensation to most everything we feel. Training yourself to tune into the physical is an excellent way to intervene before emotions become more intense or avoided altogether.

3. Start at the top of your head and work down, throughout your body. Here are a few cues to tune into: tenderness behind your eyes, heaviness in your chest, butterflies or upset stomach, tension in your shoulders, increased heart rate, sinking sensations in your abdomen, tightness. Just observe whatever physical sensations are present for you in this moment.

4. Label the specific sensations as they come over you, without merging your identity or sense of self with them: *I notice my chest is heavy* or *I am observing my heartbeat increasing* (versus *I am having a heart attack* or *Something is seriously wrong with me*). Your feelings are a compass, a guide, but they're not your identity.

5. Once you identify the sensations, see if you can label the feeling: sad, angry, joyful, afraid...Labeling and validating all of your emotions, and not just the positive ones, will help you feel safe and grounded.

6. Cultivate a warm, nonjudgmental, internal tone to welcome each emotion that comes into your awareness. Remember that feelings are neither right nor wrong; they just are what

they are. Whisper to yourself, *Welcome sadness, I am glad you're here* or *I see you, anger.* You can love yourself even when experiencing difficult emotions.

7. Your thoughts will pull you in various directions. Gently breathe in and out as you redirect your attention back to your feeling: *I see you, I am with you,* or *I want to pay attention to you.*

8. Try to observe the feelings as separate from you and as something temporary. The sensations and feelings come and go. Just as you recognize and label a feeling or sensation, it passes, only to be replaced by another and another.

You don't have to waste your precious resources avoiding yourself and your emotions. Sitting with your feelings in this way will help you to focus, connect, and enjoy your life.

Another way we become stuck in our feelings is by internalizing, or keeping them to ourselves. The more you're able to express yourself, the easier it is to process and eventually let go of difficult emotions.

Expressing Your Feelings

Small emotional fluctuations accumulate and eventually lead to bad or good moods. The more you know what you're feeling when you're feeling it, the more control you'll have over your ability to make yourself feel better. One effective way to gain a deeper sense of your emotions, and even the root causes of some of your moment-to-moment fluctuations, is through expressing yourself. There are many different ways to start expressing yourself more.

EXERCISE: Write a Letter to Your Pretend Therapist

Imagine you're going to meet with a therapist. Write out what you would like this person to know about your emotional world. Put it all down in writing, stream of consciousness, whatever thoughts come to mind. This is not a writing class, so don't worry about your grammar or spelling. This journal entry is only for you, and it's even okay to destroy the writing when you're finished. After you've completed your letter, write a response from your imagined therapist acknowledging your feelings as real while reminding you that the difficult feelings don't mean anything bad about you as a person.

There are also numerous benefits to actually talking with another human being about what we're feeling. Talking through difficult emotions helps us to better understand what we feel and helps us process our emotions more quickly. This is one reason why psychotherapy is so effective. When people hear themselves speak their private internal world aloud, they're forced to take their emotions and experiences seriously. It's not uncommon in therapy to hear a person discuss something bothersome to them and then take note: "Wow. I didn't realize how much this upsets me." In addition, the human connection typically offers another perspective and reminds us that we're not alone.

EXERCISE: Try Talking About Your Feelings

Choose a safe person to talk to about your more difficult feelings. Of course, it takes courage to be vulnerable, but the human connection will soothe your nervous system. This doesn't have to be a close friend. Relief can come from talking to a therapist or a support group, someone online, or even a neighbor.

It's important to tune into and express your feelings. When we don't slow down and notice what we're feeling, or we don't do this frequently enough, physical agitation, reactivity, and irritability increase.

Taming Your Amygdala

If you're struggling with brain fog, your amygdala may be fried. When the amygdala, the brain's fear center, goes into overdrive, chronic stress sets in, keeping you in a hypervigilant state, constantly at the ready for the other shoe to drop. As a result, you can't feel calm and grounded. You may struggle with organization, productivity, regular eating and sleeping cycles, and certainly with knowing your deeper feelings. You may even recognize that this state of being is unhealthy, but staying on high alert may feel like the only way to cope and keep yourself safe.

The amygdala is part of our less evolved reptilian brain and is responsible for a great deal of our reactions to fear and stress. It has a wide array of connections to numerous parts of the brain, allowing it to wield immense power over how we feel. Within a millisecond, before your thinking brain even knows what's happening, the amygdala can initiate a cascade of physical reactions—changes in heartbeat, surges in adrenaline, muscle tension, increases in blood pressure, perspiration, digestive changes, stomach pain, and feeling shaky or keyed up…just to name a few—that prepare us to flee a perceived threat, to fight, or to freeze in place.

For example, while you're out hiking, if your amygdala senses that you've brushed against a snake, it will swiftly steer you out of harm's way before the rest of your brain is even aware of what occurred. The amygdala's instant all-or-nothing response is effective because, after all, we don't want to waste precious time on details when we encounter a physical threat.

From an evolutionary perspective, the amygdala is our protector; it's part of how our ancestors survived and why you and I exist. The capacity to be fearful coupled with the ability to respond instantly kept the species going, and those of us who exist now carry that same neurological wiring. However, the amygdala also relies on a primitive operating system that in an emergency can completely skip over our higher-thinking cortex (LeDoux 1996). A condition of discomfort or even acute discomfort is not the same as an immediate physical threat that requires an unthinking, reflexive reaction. Unchecked, reflexive thinking can come to dominate in matters that don't require it.

Managing your life from your primitive reptilian brain means you aren't accessing the part of your brain that helps you think things through from a rational perspective. Instead, you're living like a hungry crocodile: reactive, agitated, and in fear of the next possible threat to your world. Existing this way makes it hard to focus on your larger goals and will leave you feeling depleted, irritable, and depressed.

How do you soothe your amygdala? When you're in reactive mode, you're internally agitated, and certain thought-streams play out on repeat. In particular, you think of possible threats, anticipate bad events occurring, and worry. To counter that reactivity, redirect your focus from thinking to physical relaxation. Through physical relaxation, you can bring down the reactivity of the amygdala so that you can start accessing your "upstairs brain," where you can reflect, problem solve, and understand what you're feeling on a deeper level.

EXERCISE: Calm Your Amygdala

Here are several exercises that you can do regularly to keep your check-engine light from turning on. Try them out.

Imagery: Call to mind a visual image, as you breathe in and out, to relax your nervous system. Walk through the following scene as practice:

> Close your eyes. Take a deep breath and imagine yourself standing on a beach. It's a beautiful day—not too hot, not too cold. Paint the beach scene in your mind's eye and pick up the sounds and smells: the blue sky, the warm sun on your face, the ocean waves crashing near the shore, the salty air, the feel of sand around your feet, between your toes. Picture the waves coming toward shore as you breathe in, and then as they break and recede, deeply exhale. Everything is okay in this moment; you're okay. Simply breathe in and out as you imagine this peaceful place.

Do this practice with other images that are calming to you—perhaps a mountain, meadow, or lake scene or a picture of loved ones or pets. Some of my clients have a folder on their phone with images of their kids, vacations, or even puppies that they scroll through in times of stress.

Progressive muscle relaxation: Take five or ten minutes to reduce body tension through progressive muscle relaxation. To do this, first sit comfortably. Take a deep breath in and out. Then begin to tense and release each muscle in your body: tense the muscle while breathing in for a count of three and then release the muscle tension while breathing out for a count of three. Start with tensing and releasing the muscles in your face, then move to shoulders, hands, arms, stomach, buttocks, legs, and feet. As you do this, notice the difference in sensation between tension and relaxation. Now tense your entire body while at once breathing in for a count of three and let out for a count of three. You can repeat this exercise frequently to lessen physical tension.

Deep breathing: Internal deep breathing is quick and has an enormous impact on calming the amygdala. Begin by closing your eyes.

1. Breathe in through your nose for a count of four. Observe the air entering and filling up your lungs.

2. Hold your breath for another count of four.

3. Exhale through your mouth for a count of four.

4. Hold your breath for another count of four.

5. Repeat previous steps for five minutes.

As with imagery and progressive muscle relaxation, you can use deep breathing to calm yourself when you're feeling stressed or upset. Deep breathing can be done almost anywhere and at any time.

After trying these different relaxation exercises, pick one or two and work to build them into your routine. This way you'll be adept at turning to physical relaxation when your amygdala is triggered.

Putting It All Together

As you go toward and not away from your emotional world, you will discover increased reserves for focus and memory and will experience greater pleasure in life. Next, we're going to explore certain conditions that make emotional self-awareness easier to achieve, including self-care and healthy day-to-day routines. You'll find that you're calmer and less reactive as you develop simple but effective strategies for eating, sleeping, exercise, and relaxation.

SOLUTION 5

Overcome Negative Habits with Healthy Routines

You may notice your eyes growing heavy as you begin to read about changing your habits. You may say to yourself, *I am so unmotivated, I'll never be able to do the strategies in this book.* Or, if you notice your thoughts are racing, and you're internally keyed up, you may think, *I can't pay attention to anything. This is a waste of time.* But bear with me! I promise it's worth it. My point here is that once you become aware of a physical sensation, the brain instantaneously delivers a matching emotional interpretation.

Similarly, when our lives are chronically stressful, our bodies match that experience and we can all at once feel exhausted, ungrounded, and entirely unmotivated for self-care. We struggle with over- or undersleeping, getting healthy nutrition, getting regular exercise, relaxing, and staying organized and on top of our day-to-day tasks—all of which contributes to a shortened attention span, mood swings, and memory impairment. In extreme cases, ignoring self-care for long enough can lead to developing certain medical conditions or becoming generally less functional.

When caught in the mind–body loop, it's easy to feel defeated. You check boxes and get through your days, but all the while you're depleted and mentally frazzled. When you finally pause and reflect, the state of your life and your physical self only reinforce how out of control your life and your health have become, leaving you even more cynical about your capacity to improve the situation.

As hard as it is, there's likely no other variable as effective in relieving stress and improving mood than taking consistent care of yourself.

How Negative Habits Strengthen Over Time

If you struggle with brain fog, you likely go the extra mile and then some for many aspects of your life: putting in extra-long hours for school or work, having a lack of boundaries with your family or kids, doing too much, and being too helpful and accommodating to others. Over time, others see how capable you are and come to expect you to deliver. As a result, your expectations for yourself multiply. Because you have finite time and internal resources, however, all of this extra effort leaves you with little reserves for your own self-care.

It's stressful to be constantly on for work or for the people you love, and it contributes to a tendency to treat yourself in ways that compound the stress you experience. Consider your expectations for yourself and how you may give more than is reasonable and then treat yourself in ways that reinforce your stress. See if you can identify with any of the following examples of this pattern:

Ava worked on her computer twelve-hour days Monday through Friday. A computer engineer, Ava would routinely wake up, put on the same clothes from the day before, sit on her couch, and start working. She barely stopped to eat or go outside. Every time she looked up, she was reminded of the state of her life: dishes piled up in the sink, clutter all around, a full laundry bin. She felt awful. She'd quickly detach from all that by going back into work mode. When the end of the workday came, she would finally tune into her extreme hunger. She'd order takeout and eat whatever felt good in the moment. Then she'd stay up late watching TV in an attempt to unwind. The next morning, she'd start the routine all over again.

Simon knew since high school that he wanted to be a medical doctor. When he entered medical school, he felt like all of his dreams had come true. He spent hours and hours studying and, except for a girlfriend he tried to see on weekends, had no life outside of school. When Simon's girlfriend unexpectedly broke up with him, his world came tumbling down. His grades slid and he had difficulty focusing. As the days turned into weeks and the weeks into months, Simon felt more and more like a failure, and eventually he withdrew from medical school. Without school, he had no purpose. He often found himself just staring at the inside of his refrigerator with no idea what to eat. Eventually he'd close the doors and go back to daydreaming. He'd cope by drinking and consuming pornography, just so he could feel something different from his dulled-out mood state.

Kaya's job left her feeling bored and unstimulated. Kaya worked in a doctor's office all day answering calls and scheduling appointments. Although she could be busy, there were long blocks of time with nothing to do. She felt anxious at work and spent a lot of time there worrying about various aspects of her life, making lists, and researching things she should be doing better. By the time the day was over, she felt unmotivated and empty, which became her cue to escape by spending the evening scrolling for hours through Instagram and smoking weed.

Julie got little sleep and practically lived off of caffeine. She believed that this was the only way to keep up with her high-earning career and taking care of her three kids. She'd start her morning at

5:30 a.m. with a harsh workout. She'd then drink coffee for the rest of the day as she juggled phone calls and the kids' needs. She was constantly on edge. Just hearing the chime of her phone would put her in a hypervigilant state. By the time her kids were in bed and her phone stopped buzzing, she would finally feel like she could get some real work done. Since Julie was a poor sleeper, she saw no harm in working around the clock.

Jonah took care of his wife for three years before she succumbed to cancer. When he wasn't at work, he spent most of his time with her: attending medical appointments, making sure her medications were straight, and generally helping her feel as comfortable as possible. When she died, the grief was unbearable. Jonah felt like he had no sense of self left and no will to change the situation. After work, he'd order takeout and binge on TV. He felt as if his mental faculties were declining and often wondered why he was sharper and higher functioning when his wife was alive. He told himself he was getting old and that's why he couldn't think straight.

Patterns of unhealthy self-care strengthen over time, and the longer the patterns go on, the more likely you are to turn to more destructive behaviors to escape and numb out from the stress. However, numbing out, or what I like to call "false self-care," keeps us stuck in brain fog, feeling empty and depleted.

Facing False Self-Care

Are there ways you "treat" yourself that give you a short-term high but in the long run make you feel worse? Of course, rewarding yourself with a treat is a healthy way to stay motivated. Notice, however, if you are using false self-care as a substitute for healthy coping. Here are a few examples of false self-care that will keep you stuck in brain fog:

Drugs

Alcohol

Binge-watching TV

Pornography

Compulsive video games

Risky behavior

Not leaving the house for days at a time

Bingeing on sweets

Bingeing on carbohydrates

Not eating enough

Smoking

Overuse of caffeine

Workaholism

Oversleeping

Not sleeping enough

Not doing laundry

Not paying bills

Obsessive social media or internet usage

Constant multitasking

Gambling

Obsessive attention to the news

At the heart of brain fog is almost always an overreliance on false self-care and very little true self-care. A large part of the fix is withdrawing from those habits to make room for healthy ones.

EXERCISE: Take Up True Self-Care

Here are some ideas for true self-care. Try them out:

Spend one hour a day alone without technology; dock your phone, tablet, or computer.

Take a bath.

Meditate.

Journal.

Take a walk.

Call a friend, and be present.

Start a garden.

Eat veggies and beans for a meal.

Schedule a medical physical with blood work.

Take a yoga class.

Pet an animal.

Take a child to the park, and be present.

Sit and look at the trees, the sun, the clouds.

Eat three meals today.

Organize a closet.

Clean a room.

Clean out your car.

Light incense or a scented candle.

Put your phone on silent for some part of the day.

Do one task at a time and bring your full presence to that task.

Drink a cup of tea and do nothing else but that.

Go on a road trip.

Make time to pay bills, clean up, and organize.

Add vegetables and fruits to one meal.

Read a book for fun (not for work or school).

Listen to an audiobook for fun (not for work or school).

Get a massage.

Turn toward peace by committing to one true self-care behavior every day. Doing something small and achievable brings new energy and the confidence needed to develop new habits that will have an even greater impact.

My clients are often doubtful when I talk to them about true self-care. It can feel like "a waste of time" or "a drop in the bucket" when you have many more pressing matters at hand. In reality, it's quite the opposite: without true self-care, you're likely not working at your optimal level of efficiency and not engaging your relationships with the presence they need to thrive.

Shifting Your Mood to Shift Your Life

When we feel our plate is too full, nothing reduces the overflow as swiftly as healthy self-care. We know this. However, it's also true that even though we may know what we need to do to feel better, we often can't get ourselves to do it.

For most of us, engaging healthy habits when our mood isn't onboard is a challenge. People who develop solid routines of self-care are successful for one reason: they do it in spite of themselves. In other words, they force themselves to go for that run even though a part of them would rather chill in bed with their favorite show on TV.

Now, it's important to state that if you have brain fog, you're likely to be even less motivated to take care of yourself than most people. In fact, the very concept of self-care may feel so discordant with your current mood that it sounds like torture. This is because the stress hormones you're experiencing leave you feeling like you're always treading water just to get through the day and like you have no reserves left to take care of yourself. And without healthy self-care routines, your chronic stress compounds, eventually disrupting your most essential routines of health and wellness: sleep cycles, activity levels, social interaction, exercise, food intake. The longer the disruption goes on, the more discouraging it feels to try to fix it. This leads to searching out quick fixes that match your negative mood state.

Yet despite all of this, with effort we can override unhealthy habits. People stop smoking, stop drinking, begin exercising, become parents, quit cheating, become vegetarian, and undertake any number of other life-altering routines. People change all of the time. We all have the capacity to do so. Starting healthy habits will take deliberate effort at first, but true self-care pays off exponentially.

EXERCISE: Imagine Life If You Took Better Care of Yourself

Increase your motivation for healthy habits by making a list in your notebook of what awaits you with true self-care. How will your stress level change if you're able to exercise more regularly or get more sleep? How would you feel about yourself if you took some time for yourself most days?

Sticking to new habits will be hard at first, but doing so means you're moving toward peace, structure, clarity, and competence.

How Healthy Self-Care Impacts the Brain

Dopamine is a powerful chemical in the brain that makes us feel good. Dopamine can energize healthy behaviors—for example, dopamine may be increasing as you read this chapter, encouraging you to take on some new challenges and habits. Unfortunately, dopamine reward pathways also light up for unhealthy habits and behaviors. At the extreme, drug use gives an immediate, hugely intoxicating surge. And though not as intense as drugs, other habits like pornography, gambling, bingeing on food and video games, drinking alcohol, and bingeing on TV all provide dopamine surges. Even clicking on apps, social media, and websites activates the dopamine reward center in the brain. And as hard and all-consuming as workaholism can be, it too provides a dopamine hit.

Similar to a kid at a carnival whose single focus is more...more rides, more candy, more popcorn, more thrills, more fun...when our dopamine reward center is triggered, instead of thinking rationally about how we'll feel when the hit wears off, we keep searching out another hit.

The pattern of going back and forth between hit and withdrawal is, of course, self-defeating. Anyone who's wasted a few hours scrolling social media or searching for the end of the internet will ask themselves, *Why did I waste my time like that?* In fact, you likely scold yourself on a regular basis for not taking on healthier day-to-day habits. You know better. Still, you can't get yourself to do what needs to be done. When you decide to forgo healthy eating and exercise in favor of junk food and the couch, you're not thinking about how defeated your future self will feel.

The good news is that the same neuronal pathways that light up for unhealthy behaviors can light up for healthy habits. In fact, dopamine also responds to situations where we achieve and master our circumstances. So making a plan right now to increase healthy self-care behaviors, and sticking to that plan, will give you a dopamine hit on a day-to-day basis. And as these healthy behaviors turn into habits, you'll increase other pleasure-producing chemicals in your brain. These chemicals foster a sense of well-being and positive feelings of belonging and counter depression and irritability.

Taking care of your brain through nutrition, sleep, and exercise and relaxation routines sets the stage for improved memory and focus. Solid self-care is the foundation of the house, so to speak, and produces the optimal conditions to reduce brain fog.

Which Habits Need Your Focus

As you take in this chapter, you may be overwhelmed by what I'm asking of you. You may be doubting yourself, wondering if you can change, second-guessing what is good health, or just expecting more of yourself than is reasonable. Making a change can come to feel so hard that you lose motivation, give up, and go back to your old ways of coping.

In truth, it's the simple habits that make all of the difference. You don't have to run a marathon or eat veggies 24/7 to start adding more health to your life. Here's a list of good daily habits that will start a new pattern and give your brain the nurturance it needs.

Nutrition

Food is, of course, one of the great pleasures of life, but it's also fuel for brain health. For example, research shows that consuming flavonoids, a chemical found in plants, prevents cognitive decline (Yeh et al. 2021). People who consume at least one-half serving a day of flavonoids, including apples, oranges, peppers, strawberries, and pears, are less likely to report confusion and memory loss. And what's really inspiring about this research is that people who started consuming flavonoids only more recently showed the same cognitive benefits as people who had been consuming them for twenty years. Similarly, the Brain Health Food Guide, developed by scientists and nutritionists at the Center for Aging and Brain Health Innovation, outlines nutritional habits that aim to prevent cognitive decline, including memory and thinking (Baycrest 2017). Adopting the types of eating patterns recommended in the Brain Health Food Guide is associated with increased cognitive functioning, including decreased memory loss (Smith et al. 2010; Valls-Pedret et al. 2015).

EXERCISE: Easy Ideas for Healthy Eating

Start to improve your nutrition today by implementing these ideas for healthy eating:

Consume a variety of fruits and vegetables.

Limit red meat.

Focus more on overall healthy eating as opposed to one particular "superfood."

Drink more water.

Eat leafy greens daily, including kale, spinach, and lettuce.

Eat beans and legumes several times a week.

Eat berries several times a week.

Snack on cut-up veggies, fruits, and low-fat yogurt.

Increase cruciferous vegetables (e.g., broccoli, cauliflower, brussels sprouts, cabbage, kale).

Eat fish and nuts a few times a week.

Rid your pantry of processed foods and sugar.

Reduce caffeinated beverages.

When chronically stressed, it's easy to become overly dependent on stimulants, like coffee and high-energy drinks, because they help us to feel more alert and focused. Overreliance on stimulants however, will keep you feeling physically keyed up and on edge, which will only intensify brain fog. Consider removing all caffeine from your diet. If that feels undoable, cut your caffeine intake back by half and work down from there.

With brain fog, it's a good idea to abstain from nicotine and alcohol. Both nicotine and alcohol light up the dopamine reward centers of our brain. As the high diminishes, it's easy to become more stressed and anxious so that you seek another hit to reduce these symptoms. In fact, you can revolve your entire day around when you'll have your next cigarette, or you can start living for your nightly cocktail or glass of wine. If you're working to overcome brain fog, take thirty days off from drinking and smoking. For some people, this change alone starts a new pattern of increased energy and focus.

Sleep

Getting good sleep is central to good physical and mental health, and sleep disruption is on the rise. Fifty- to seventy-million adults in the United States meet the criteria for a sleep disorder, with the most common disorder being insomnia (American Sleep Association 2022). When we don't get enough sleep or get too much sleep, we're likely to have other struggles as well, including depression, anxiety, irritability, and cognitive impairment. The answer is to learn to treat sleep as a special and sacred part of your existence, adopting a calming nightly routine that will prepare your brain for sleep.

When healthy sleep hygiene is practiced regularly, our nightly sleep routine tells the brain that it's time to start unwinding. The key is to follow the same habit consistently. Eventually, you'll need only to start your routine to feel more physically at ease and sleepy.

EXERCISE: Get into a Calming Bedtime Routine

Start your bedtime habit at the same time most nights, ideally one hour before going to bed. Turn off electronic devices including phone, television, and computer. Take a warm shower, change into your comfortable clothing, and drink something warm without caffeine. Lie down restfully, and do a relaxing or soothing activity (read, knit, breathe). When you begin to feel sleepy, turn off the lights and do a relaxation exercise: practice progressive muscle relaxation or visualize relaxing imagery. Note: If you can't sleep, instead of asking yourself, *Why can't I sleep?!* tell yourself, *It's okay if I don't fall asleep. At least I'm resting.* If waking continues, try progressive muscle relaxation again, with the lights off. Put your clock out of sight, because it's best to focus just on relaxing your body, even if you can't sleep.

Rise at the same time each morning; generally seven to nine hours of sleep is plenty. Too much and too little sleep will mess up your sleep hygiene. Developing a good bedtime routine will help you get a better night's sleep. Don't try to compensate for a poor night's sleep by dozing during the day or by going to bed before your usual time.

Exercise

The benefits of exercise are numerous. If you struggle with brain fog, your perfectionistic thinking may be getting in the way. You may tell yourself, *A walk is hardly worth it. I should be in the gym,* and so end up doing nothing because a walk doesn't feel like good-enough exercise. Hear me on this. A walk is good enough!

What's important is to make a realistic exercise goal. If this means walking, yoga, or hiking, or even strolling around the mall, that's okay. Try to pick something that's more than what you're currently doing. After two weeks, increase the intensity of the exercise or the time you spend exercising. So, for example, maybe you start by walking every day for fifteen minutes. After two weeks, increase the time to thirty minutes. Then consider a light jog for fifteen minutes and a walk for the second half, and so on. Building fitness means increasing ever so slightly the time spent or the intensity of the exercise.

Consider something positive to do along with exercise that will deliver some kind of reward…a dopamine hit…that keeps you coming back for more. Maybe it's exercising with a friend or participating in a group fitness class. For me, it was starting Orange Theory Fitness after not exercising so regularly. Each time I see that orange light illuminate, I feel like I've won the lottery!

Important note: Check with your medical doctor that your exercise routine is safe for your physical condition.

Take a moment now to write in your journal what your exercise goal is. Each time you exercise, you'll increase your body's endorphins, your cognitive functioning, and your sense of believing in yourself.

Relaxation

Start increasing true relaxation time by simply decreasing the time you spend on false self-care, or numbing-out habits. Instead of eliminating false self-care entirely, start by reducing it by half and work up from there. For example, if you usually scroll Instagram for two hours most days, set a timer and stop scrolling after an hour. Then take the next hour to train yourself to relax in a different way.

There are many healthy ways to relax, and later in this book we're going to get into the specifics of mindfulness training, which will help you learn to appreciate just being in the moment. One way to give your brain a jump-start in this department is to simply commit most days to some time outdoors in nature. Research shows that spending time in nature increases working memory, cognitive flexibility, and attention (Schertz and Berman 2019). Also, if your brain is used to being keyed up, nature offers enough distraction to allow you to be at ease within yourself without feeling too bored or too hyped up. Bring your full attention to what's around you—the color of the sky, the leaves, the grass, the smells, and the sounds you hear. This single practice will clear your brain and help ground you in your body.

Putting It All Together

It likely feels overwhelming to make some of the changes we've explored in this chapter. Remember: any change you make, no matter how small, is the start to a new pattern of physical health and wellness. Exercise, relaxation, and healthy sleeping and eating reinforce each other. So if you exercise and get outside during the day, you're more likely to sleep better at night, and if you sleep well, you're more likely to be motivated to exercise the next day. Healthy habits are necessary to begin taking your understanding of brain fog to the next level. Let's start looking at how negative thinking contributes to the fog and what to do about it.

SOLUTION 6

Overcome Feeling Defeated with Your Thinking Brain

For a moment, I'd like you to think of nothing at all, to clear your brain completely of thoughts—absolutely no thinking allowed. If you find yourself thinking, immediately turn off your brain and stop thinking! At the end of this paragraph, stop reading and sit still for a minute, long enough to turn off your thoughts...and afterward start reading again.

Can't do it, right? Our thoughts are on a continuous and never-ending loop. We have thoughts about everything we do or don't do, and even when we're doing one thing, we can have an entirely different train of thought operating in the background. Our thinking brain defines our identity and moment-to-moment experience of ourselves, impacting how capable, happy, and present we feel. When our thoughts are productive and supportive, we thrive; when our thoughts are negative and defeating, we shrink.

Consider a beautiful, clear pond that is slowly overtaken by algae. As the algae grows, it turns the clear water a murky green and comes to define the pond. Similarly, when our thoughts play on a consistently defeating pitch, we come to see our own potential for happiness and achievement as diminished. This dark place becomes our reality and we forget that unclouded, clear thinking is possible.

Conditions differ from person to person, but they typically consist of some combination of constant worry thoughts or *what if* possibilities—catastrophizing or imagining worst-case scenarios...obsessing about events or possibilities that in reality you can't control or be certain of...mentally replaying events from the past...or constant planning for the future. One reason you may stay stuck in this place is that you simply don't know another way to think. And the longer you've existed in this state, the harder it is to believe you can change.

Just as the eutrophication of a small pond can be reversed, with deliberate effort, you can bring your negative thoughts under control and reestablish your sense of contentment, drive, and belief in yourself. The key is to slow down and observe your thoughts so that you can effectively manage their content.

Observing Thoughts versus Getting Caught Up

Struggling with brain fog means you're likely bombarded by your thoughts, making it hard to be effective or to focus on what's most relevant to your life and functioning. Thoughts can be so overwhelming that we don't even recognize that it is self-defeating to ask ourselves to believe and follow every thought that comes into our awareness. If you give yourself the space to *observe* your thoughts, however, your thoughts will move in and out of focus with greater ease; this is the nature of the mind.

EXERCISE: Practice Observing Your Thoughts

Here is an exercise to start observing instead of getting caught up in your thoughts:

1. Take a moment to slow down and sit with yourself. Take a few deep breaths in and out.

2. Picture a part of you as separate from your thoughts, an observer whose only job is awareness.

3. Nothing else is required of you in this moment but to be aware of what you're thinking. Watch the thoughts as they come in and out of your consciousness.

4. Consider these questions as you observe your mind: do your thoughts bounce from topic to topic, or are there a few main categories your mind is dwelling on? Are the thoughts anxious, sad, angry, or planning oriented? All of the above?

5. See if you can label categories of thoughts as they come into your awareness, without actually getting caught up in them, such as "worry thoughts entering..." "health thoughts exiting..." "planning thoughts entering..." Other common categories of thoughts can be work, school, self-esteem, relationships, and family.

6. You are not your thoughts. You're actually the one observing your thoughts enter and exit your conscious awareness. Remind yourself of this by saying out loud that you're observing or noticing a particular thought. So instead of saying, "Oh my God! I am going to blow that work project," say, "I am noticing I am having the thought that I am going to blow the work project."

7. Without judgment or criticism, just let your thoughts know you're aware: "I see you, worry thoughts" or "I see you, self-esteem thoughts."

8. When you're ready, end the exercise.

Observing your thoughts allows thoughts to naturally enter and exit your awareness. It helps you see how your thoughts pass when you stop actively trying to control them.

This simple practice in observing your thoughts can help you begin to step back from self-defeating thinking.

Your Thinking Brain

The cerebral cortex, your "upstairs brain," is the extremely complex outer area that allows you to reflect, problem solve, plan, imagine, anticipate, and rationalize. It's the seat of your awareness and consciousness. Each and every moment of your existence, your cortex is generating thoughts: some creative, some random, some reality based, some entirely irrelevant. In optimal circumstances, we naturally sort our thoughts all the while remaining focused and present.

When you're chronically stressed, however, a continual contingency analysis takes hold that has the potential to dominate brain space, choking out the life you want to live. Before you know it, you're spinning in your head—wondering, thinking, guessing, imagining— *If I go there, then that will happen…If I get that result, then this will happen…What will happen if I say that?…If I don't do that, then this will happen.* Much of your conscious life becomes preoccupied by possibilities that in reality may not be as risky to you as they seem in your mind.

In the best light, we wake up thinking: *Kids are going to day camp. I need to wake them up…and it looks like rain so will pack raincoats. Take shower, drop off dog, and then head to work. Pick up groceries for dinner* on the way home. In the worst case, it's *Ugh, the kids are still asleep, and we're going to be late. I don't have time to find the raincoats...I am a terrible parent. What will we eat tonight? How can I possibly plan dinner when I have to write this brief for work! I am so screwed. If I don't write this brief, I am going to look like an idiot at the meeting today.* Suddenly this stressed-out thinking has us sped up to such an extent that we can't effectively deal with the situation at hand.

Again, the cortex makes interpretations and draws conclusions about the circumstances in your life. When operating at full efficiency, we're better able to find what's useful and discard that which is inaccurate or irrelevant in our thoughts. When stressed, however, we become consumed by upsetting interpretations, many of which harm our functioning or are entirely untrue.

Grant's Story

Grant had two wonderful dates with a new romantic prospect. After the second time seeing her, he felt confident that his date had fun too and sent a text saying how much he'd enjoyed the evening. When he didn't hear back right away, however, he went into an overthinking spiral. Though he initially believed his date had gone well, Grant quickly succumbed to an old habit of dealing with uncertainty by consuming himself with self-evaluation and what he may have done wrong. He concluded that some of his comments must have been offensive and that he was boring and not funny enough. Then, after a few days of anxiously thinking and drawing negative conclusions, Grant received a warm text from his date, who had been out of town and unable to respond.

While Grant, upon receiving no early response to his text, may have thought his negative interpretation was correct, he actually descended into an unnecessarily anxious spiral. We humans like to take our thoughts extremely seriously but, in reality, the cortex is not an unbiased reporter and makes all kinds of subjective interpretations. Oftentimes the cortex generates meaning and significance that isn't really there but is more reflective of our past experiences, trauma, or learning history. Once you see the truth about the cortex, you can let a whole bunch of thoughts go.

Even though you recognize that thinking about something doesn't make it true, you may still find yourself caught up in believing everything your thinking brain tells you. This is because, as you learned in chapter 2, repeated experiences over time trigger the same patterns of neuronal activity. This is why anxious and stressful thoughts are so hard to release; the more you engage certain worry or stressful thoughts, the easier it becomes to activate these same thought-streams in the future. It may seem to you that since the thought keeps coming up, there must be something important to fix or to worry about. Actually, the stressed-out thoughts keep coming up not because they're true necessarily but because your brain has associated certain events with certain thought-streams for so long, these worried thinking loops have become sewn into the fabric of your mind.

Defeating and stressed-out thought patterns can be effectively challenged, however, with deliberate effort.

Challenging Your Thoughts

If you're reading this book, somewhere along the way to brain fog you likely adopted the belief that to keep afloat, stay sharp, and not make mistakes, you have to stress out and overthink (*If I am not anticipating all of the possible things that could go wrong, they'll go wrong!*). Anticipating worst-case scenarios doesn't make them any less likely to happen, however, and it depletes emotional reserves, in effect rendering you helpless when adversity does strike.

Consider for a moment what would happen if each time a thought popped into your head, you just allowed yourself to treat it as a fact. For example, you receive a low test score, and you determine, *I'm not getting into a decent college.* Or you experience some turbulence on a flight, and you determine, *We're going down!* Or your car breaks down on your way to an important meeting, and you determine, *Bad stuff always happen to me.* Or you are left out of a social event, and you determine, *Everyone is turning on me.* Catastrophic thinking puts us into panic mode, making it hard to effectively manage a given situation. Luckily, there are some effective strategies to counter this kind of thinking.

If you tend to catastrophize, try to catch yourself when you're drawing a negative conclusion and consider your setback from a more realistic perspective. I'm not asking you to lie to yourself but rather to consider the most likely, or the most reasonable, possibility in the worrisome scenario. A good way to get there is to consider the worst-case scenario that your cortex is generating, and then call to mind a best-case scenario. You'll find the most likely scenario is usually somewhere in the middle.

For example, say your car breaks down on the way to an important work meeting where you're expected to make a presentation. Ask yourself these questions to explore the worst, best, and most likely scenarios:

1. How bad could the outcome be? Example: *I'll get fired.*

2. Is there a conceivable way in which this ends well? Example: *The meeting gets delayed or canceled, and they don't even know I'm late to work.*

3. What is the most likely outcome? Example: *I'll be late and do my best to salvage the situation. But I can't change the situation; it is what it is.*

Stepping back in this way can help you come to a more realistic understanding of the situation, which will lower your stress level and help you manage better.

Another way to get perspective on catastrophic thinking is to consider any evidence proving that your stress or worry is justified and any evidence disproving that it's justified. For example, imagine you are at a social event and you find yourself all alone and unable to contribute to any conversation. Suddenly, you feel like you don't belong and you worry you're unlikable. Stop and consider whether this concern is justified by looking at the evidence for and against it:

Evidence proving your unlikability: *I am at a social event and no one is talking to me. I feel like no one even knows I am here.*

Evidence disproving your unlikability: *I have friends outside of this specific situation who like me and want to see me. I'm more outgoing in smaller social situations.*

Considering the evidence is another way to gain some perspective when you find yourself coming to a worst-case conclusion.

We all face hardships; it's just part of the human condition. And all the worry in the world won't prevent you from struggling from time to time. Replace automatic conclusions that keep you spinning with ones that help you feel calmer, more at peace, and more in control. Of course, at first your new conclusions will seem unreasonably positive and even phony to you, but try to just go with them in spite of yourself. Your new conclusions will seem more and more natural over time.

Common Types of Faulty Conclusions

Try to be aware of when you're drawing a faulty conclusion out of habit. Work to replace faulty conclusions with thoughts that help you to adaptively cope and feel more at peace. Here are some common kinds of conclusions that we tend to draw automatically but, in reality, are faulty and only increase stress; this list includes examples paired with more adaptive coping thoughts.

Catastrophic thinking. Something bad happens, and you assume the worst-case scenario will result. (See previous example of car breaking down on the way to work.)

Making the uncertain certain. Instead of accepting uncertainty as an inevitable part of life, your brain searches for answers. Example: *I have to get my annual blood work. What if I have cancer?* Coping thought: *It's not my job to know things that I can't possibly know.* Or *Until I find out the facts, I am turning this over to the universe.* Or *I'd rather accept uncertainty than spin my wheels and get nowhere.*

Black-and-white thinking. No room for the in-between. Example: *I started this project late; I'll never be able to complete it.* Coping thought: *I've started projects late before, and it's been okay. I'll do the best I can with the time I have.*

Judging. Applying judgments to yourself and others that are sweeping statements of worth, like *I'm a loser* or *I'm boring.* Example: *I can't understand this computer program for work. I'm always the dumb one in the group.* Coping thought: *Everyone has their strengths and weaknesses. I can learn the specific skills for this program and improve over time.*

Unrealistic expectations. You have a perfectionistic mindset that you and perhaps others must act in certain ways or do very specific things or else there will be irreparable consequences. Example: *I wanted to get back from my work trip, go to my friend's party, and then make a volunteer event, and now I'm delayed. I'm failing at everything.* Coping thought: *It's okay to say no; it's impossible to do everything and be a healthy person.*

Superstitious thinking. This is when you convince yourself that stressing out and worrying is a way to stay safe when in reality all it does is make you more anxious, so it's harder to think straight and take care of your situation. Example: *If I don't worry about…then the worst will happen.* Coping thought: *Worrying won't make it more or less likely to occur.*

Extrapolating out across time. This is when you take one bad moment or data point and conclude that you'll always struggle in the same way. Example: *People keep breaking up with me. I'll never find a partner.* Coping thought: *Each relationship is a new start with new potential.* Here's another example: *That always happens to me.* Coping thought: *I can think of some times when that hasn't happened to me.* And this example: *I was passed up for the promotion. I am stuck in this job forever.* Coping thought: *I'll find out what I'm missing so I can have a better case for a promotion going forward.*

Self-doubt. This is when you discount your effectiveness for dealing with difficult events or unwanted future scenarios. Example: *If that happened, I would die.* Coping thought: *If that happens, I'll work through it. There will be a path forward.* Or this example: *I couldn't handle that.* Coping thought: *I don't want that to happen, but if it does, I am capable of managing far more than I give myself credit for.*

Identifying faulty conclusions will help bring down your stress and help you see life and hardships in a more realistic light.

Ruminating versus Problem Solving

Rumination is when we internalize our stress through repetitively thinking about all the reasons or causes for why we're upset or stressed. For example, you receive negative feedback at work. You find this hard to hear, and you don't immediately know what to do about the situation. Your mind starts racing with the possibilities of what other negative opinions your colleagues have of you, what they thought about various projects you did in the past, what work you won't get going forward, and what must be wrong with you from childhood that caused you to have these work problems. Before you know it, you're flooded by stress hormones and your heart is racing. You can't think straight, and when your child or spouse asks you for something simple, you explode. Now you have even more things to ruminate about, because you feel guilty about your parenting or not being a good partner.

In the moment, rumination can feel like problem solving, because while we're overthinking, we feel less helpless over our circumstances: *If I just think about this enough, I'll figure it out and solve the problem.* In reality, ruminating is passive and so keeps us stuck.

Try to differentiate ruminating from healthy problem solving. The latter means considering what's upsetting you and then running through the possible actions you can take to mitigate or solve the problem at hand. If you can't identify any action to take but continue to think about it, you're spinning your wheels, wasting precious energy, and going nowhere. When this is the case, practice acceptance, turning the distress over to the universe. Remember: you can't be perfect, and you can't control everything.

EXERCISE: Problem Solving

In a notebook or your journal, make two columns. In the first column list your current repetitive stress points (the issues and events you think about on a regular basis). In the second column, write out a few behavioral actions you could do that would help you cope with or mitigate each of those stress points. Now, circle at least one behavioral action for each of your stress points and commit to taking that action.

When there's no action you can take to change or improve things, turn your stressor over to the universe. Learning to accept that there will always be things out of your control brings peace.

Learning to Exist with Clarity

Similar to slime consuming a pond, stressed-out, negative, defeating thought-streams have the capacity to take over your entire awareness. Instead, you can learn to exist with clarity, interrupting your thought patterns so that you can challenge and let go of unhelpful thoughts.

EXERCISE: Journal Your Thoughts

Here's an exercise that uses the tools in this chapter to help you slow down and consider your thoughts more carefully.

1. Write down the stressful situation, images, events, or interaction that you're thinking about. Example: *I am worried that I have a serious health issue and will have to undergo many medical*

tests. My life will be painful and shortened as a result of this condition. I have a lot of responsibility, and there's no way I can deal with this too. Everything is going to be awful, and I'm going to fall apart.

2. Consider any faulty conclusions you may be forming. Example: *Catastrophic thinking, self-doubt.*

3. How stressed out are you by this thought on a 1 to 10 scale, with 10 being the most intense? Example: *8.*

4. Consider any evidence proving that your stress or worry is justified: Example: *I have a regular migraine every afternoon. At least once a day, I am dizzy and unsteady on my feet. I am already behind with work and family commitments and can't handle another thing.*

5. Consider any evidence disproving that your stress or worry is justified. Example: *When I don't have the migraine, I am able to work and feel okay. I had some blood work six months ago, and everything was okay. I am also very stressed, so that might be impacting me physically.*

6. How bad could the outcome be? Example: *I have a condition that is going to kill me, and I won't be able to function or take care of my family.*

7. Is there an ideal scenario? Example: *Nothing is wrong, and I am in perfect health.*

8. What is the most likely or realistic possibility? Example: *I haven't taken good care of myself lately. I may have some symptoms that I need to address, but they likely aren't life-threatening.*

9. Replace faulty conclusions with more adaptive coping thoughts. Example: *I can cope with hard things. Some good might come from taking the symptoms seriously; it will motivate me to take action to decrease my stress and get on a healthier path.*

10. Is there anything you can do to mitigate the situation, and can you turn the rest over to the universe? Example: *Set up appointment with my general medical doctor. Get more sleep. Take a walk every day.*

11. How stressed are you now after completing this exercise, on a 1 to 10 scale? Example: 6.

Use this list of questions to decrease your stress level. Even if your stress decreases by a small amount, this is significant and shows you can make yourself feel better. Each time you catch yourself in self-defeating or faulty thinking, remind yourself of your realistic scenario and positive coping statement. In this case, you may tell yourself, *I see your worried thought. I can cope with this, and maybe it's a way to start a healthier lifestyle in which I feel more in control and better.*

The next time you notice your brain spinning, slow down and journal your thoughts in your notebook.

Putting It All Together

Maintaining clarity while coping with life's inevitable hardships means teaching yourself to slow down so that you can observe your automatic thought-streams and replace defeating thoughts with realistic thoughts. Also, active problem solving and taking action significantly decrease repetitive rumination. As you deliberately engage new ways of handling your thoughts, you'll spend less time in a rabbit hole of worry and more time in the here and now. Let's turn to another effective tool for increased presence: mindfulness.

SOLUTION 7

Overcome Irritability and Stress with Mindfulness and Acceptance

While driving to work, you mentally rehearse what needs to be accomplished when you arrive. As thoughts flood in, your heart rate increases, you feel physically agitated, and you're anxious to get in and begin. You've got a death grip on your steering wheel, but in your mind you're typing on your keyboard, firmly planted at your desk. A driver cuts you off, and suddenly you're caught up in a snarl of traffic. Your body's stress response is fully engaged; you're even more tense and jittery, only now you're also angry. You honk the horn, and a new narrative takes hold centering on what's wrong with the driver in front of you (*What a jerk! Why did he cut me off?!?*) and how you don't deserve to be treated this way. Your shoulders are hunched up. As you dart in and out of traffic, your thoughts race with other injustices you've experienced.

Brain fog and distractibility go hand in hand. Your mind is perpetually elsewhere. The context changes depending on what's going on in your life, but this type of constantly replaying the past or anticipating the future comes to define your experience of yourself. You feel always behind and struggle with focus because you're never fully engaged with what's happening for you in the here and now.

Mindfulness and acceptance practices are ways to interrupt the brain's tendency to bounce between the past and the future. Mindfulness means adopting a calm, warm, present-focused attention to all of your experiences—both joy and suffering—and it means increased intimacy and ease within yourself. This internal ease extends outward to sharpen focus, concentration, compassion, and patience with others.

Luna's Story

Luna woke up like most days in a pretty good mood. She had her coffee as she perused the news, but when her son and her partner came in for breakfast, she would become irritated. Luna deeply loved her son and her partner and would later beat herself up for not being kinder and gentler. She'd will herself to do better.

Luna always ran herself hard. As a mother and a business owner, she literally could never get a break. Luna described her day-to-day experience as being like attempting to solve a never-ending math problem with a loudly barking dog in the background, kids running in and out, and a meal burning on the stove while her phone blows up.

A medical diagnosis put Luna's stress on a whole new level. For months, Luna had struggled with fatigue and weakness but assumed it was due to her hectic work–home life imbalance. Then she started experiencing pain in her back, difficulty walking, and dizziness. She finally went to a doctor and was diagnosed with multiple sclerosis. This was too much to bear, and Luna couldn't accept it. She continued the only way she knew how; she worked even harder to keep afloat. The problem was her body was exhausted and unable to keep up with her mind.

Luna beat herself up: What's wrong with me that I can't get my responsibilities completed? I have to work harder. This stressed-out thinking only intensified Luna's pattern of feeling irritable, physically agitated, and tense. Eventually, Luna was *hospitalized with exhaustion and dehydration. It was at this point, seeing the worry in her partner's eyes, that she came to see me.*

Luna began to engage in mindfulness and acceptance practices. As she put it, "What else can I do at this point?" She stopped ignoring her body and instead tuned into her breath, body tension, and other physical sensations. This little bit of space made a big difference, and Luna felt more physically at ease and calm.

Luna became aware that she worked in overdrive to compensate for life's unpredictability. Instead of trying so hard to fix things, she allowed space "just for me to be me."

Like Luna, in an effort to feel safe, to feel at peace, and to be successful, you may work to control each and every aspect of your life. When you push yourself to such an unreasonable extent, little is left for your inner self. Naturally you become irritable, distracted, and disconnected from the present moment. If this sounds like you, this next exercise will help you reconnect with the here and now.

EXERCISE: Quick Mindfulness Exercise

Sit comfortably in a quiet place. Take a deep breath in and out. Now systematically go through each of your senses and bring your full attention to what you notice and observe. In this moment, what do you see...what do you hear...what do you feel on your skin...what do you taste in your mouth? Each time your thoughts wander, redirect your focus back to your senses as you breathe in and out.

You'll learn more about how to use mindfulness later in this chapter. But first let's get into a related topic that's equally important: accepting what you can't change.

From Suffering to Acceptance

Suffering typically consists of two components. The first component is the setback itself: the loss, the work challenge, the medical diagnosis, the global event, the child suffering, the toxic relationship, the trauma, the deadline, the irritability, the fatigue, the speeding ticket, the exclusion, the relationship ending, the failure, the job loss, the lack of income. The second component is what we inflict upon ourselves regarding the setback: the *This can't be happening*, the *What did I do wrong to make this happen?*, the "*I have to fix this or else*, the *How could they hurt me like this?*, or the *I am inadequate. That's why this happened to me.* And the second component is also the obsessive and enduring sense that because of the original setback, we can't and won't be okay, or won't be okay until it's entirely resolved. This secondary component is self-defeating and the only part of suffering we can influence.

We can influence this part of suffering through acceptance. Through acceptance, we see situations wholeheartedly as they are, as they truly exist. We cease willing things to be different and instead make choices about how best to mitigate the limits of our situation. We operate from a self-possessed presence; we are courageous and acknowledge what's real and true so that we don't waste our precious energy spinning our wheels on what can't be changed.

When not in a place of acceptance, we operate from anger and fear and feel ill at ease and insecure.

Imagine for a moment you've booked the vacation of your dreams. You're going to Tahiti, and you've splurged for one of those overwater bungalows, the kind with the transparent flooring, where you can see the fish as they swim directly below you, and with the luxurious deck where you can jump into your private bit of ocean whenever you like. When you arrive to your resort, you're beyond excited. The hotel staff usher you by golf cart to...oh no...not your overwater bungalow but a regular hotel room on the property. You immediately protest, "There's been a mistake. I booked an overwater bungalow!" You reserved it months ago and show your confirmation email as evidence. The reality is that no overwater bungalows are available, because the resort is at capacity, so you won't be accommodated on this trip. The hotel staff are very sorry. Your rage is palpable: *This can't happen. It's not right. This is unacceptable!* You've saved for months for this trip. It feels like they're taking from you what you deserve, what you earned, and that they're hurting you and doing it willfully.

At this point, you have a choice: you can spiral into an obsessive universe, even hire an attorney, and book a flight back home, or you can come to your senses. Yes, the situation is not as you wish, but you can still enjoy the other components of the vacation: the crystal blue water, the warm sunshine, the mountains, the beautiful vegetation. And as you take a closer look at your room, it's actually quite lovely and, although it doesn't have the private bit of ocean you wish for, it does have a nice vista. You find a way to accept the situation, so you can enjoy your precious time on the island of Tahiti.

We only have one life and a finite amount of time. Wasting energy with self-critical thinking, anger at others, and obsessive working to improve things that we aren't in control of means we miss out on other healthy parts of life and the joys of this moment.

For Luna, this meant accepting that she could no longer work at the same level, because of her medical condition. She accepted that if she didn't change her expectations for herself, she would become sicker. Luna fought this reality for a long time, and it brought on grief. But as she accepted it, she saw an opening for other choices. Luna eventually worked less, spent more time with loved ones, and became better rested and stronger. Over time, she found herself more present, focused, and alive than she had been in years.

Acceptance doesn't mean you have to like or want whatever it is that's plaguing you. Acceptance doesn't mean you failed or that you're giving up or giving in. It means recognizing that some hardships simply are what they are and can't be fixed and controlled, and that willing yourself to try to fix that which is out of your control is a losing battle that merely intensifies your suffering.

The moment we accept our setbacks and vulnerabilities as they are, we gain true power and choice over our circumstances. Here are some other examples:

When John accepted that his mother had a serious drug addition, he recognized how much of his energy he wasted on trying to get her to see that she needed help. John started to put his energy toward himself, his own goals and happiness.

When Mia accepted her husband's recurrent pattern of unfaithfulness, she stopped making excuses for him and felt less burned-out as a result. This renewed energy helped her to consider whether she wanted to enter couples therapy or start the divorce process.

When Judah accepted that his son Liam had a learning disability, he stopped getting irritated with him for not trying hard enough and being lazy; they became closer as a result.

When Lindsey accepted she couldn't do everything perfectly, she stopped beating herself up and felt more at ease and relaxed.

When Bess accepted that her neck pain wasn't going away, she went to the doctor and underwent surgery that eventually made her more physically comfortable.

Seeing things as they really are means we can make choices that help us feel better in the here and now.

EXERCISE: Practice Acceptance

Think of some examples in your own life where you could improve your situation through acceptance. In the case of brain fog, you may find these ideas helpful:

Accepting that you have limits

Accepting you're imperfect

Accepting help

Accepting that the way you're operating isn't working

Consider what you need to accept, and then say out loud: "It is what it is." As you do this, even for a moment, see if you can feel relief in letting go, not having to work so hard to hide or cover up what is clearly true and real. Perhaps other choices will become possible.

Now that you've explored acceptance of what you can't change, let's return to mindfulness to see how you can practice being in the here and now.

Exploring Mindfulness

We can be surrounded by beauty, the change of seasons, a certain slant of light, or a place where blue sky and green trees brilliantly complement one another, and yet all is lost on us. In the same way, our life experiences—raising children, meaningful work, self-understanding, loving friends and family, deep belly laughs, closeness with a romantic partner, gratitude—all may come and go, not fully experienced.

There's no force as strong as time. Our children grow up, cherished others pass away, nature changes, jobs end, people move, physical limitations set in, the days ahead of us are fewer. We'll have regrets and wish for more one day: likely not more money or more prestige or more work, but more time with others, more time appreciating, more time seeing this world with wonder.

To wake up over and over again to the experience of the present moment is mindfulness. Based on ancient spiritual practices, mindfulness is a training of the mind to be aware of what's occurring in the here and now. Mindfulness training decreases depression, rumination, and stress and improves psychological well-being and relationship satisfaction as well as focus and attention (Davis and Hayes 2011). Mindfulness practices are effectively used in a variety of settings, including schools, the workplace, parenting practices, and even in helping with medical illnesses and physical pain.

The act of mindfulness is essentially monitoring yourself for when your mind wanders and then gently redirecting your focus to what's occurring in the here and now. This could be as simple as focusing on your breath, sounds around you, physical sensations in your body, or what a person is sharing with you in that moment. The goal is *not* to never be distracted, an impossible task! The goal is to wake up your consciousness, to build awareness of distracting thoughts, and to let go of them and come back to the here and now.

When you observe your pleasurable experiences as they're occurring, whether this be connecting with a loving partner or a friend, eating a meal, observing nature, or sitting with your pet, you encode the experience, and this helps you feel full, grateful, and connected. These experiences are protective for when inevitable hardships occur.

When you don't deny but observe your more painful experiences, the experiences pass with less suffering and you become aware of what you need going forward. And when you apply a mindful awareness to your relationships, when you're really with the person in front of you and you feel connected to yourself at the same time, the relationship will improve and compassion and connectivity will grow. When you're truly embodied as you walk, talk, view nature, and hear and see what's around you, peace comes.

Becoming mindful means employing a friendly, interested, and caring attention to yourself and all of your experiences. Here are the components of this practice:

Being Nonjudgmental

We often use judgments—right or wrong, good or bad, fair or unfair, like or dislike—as a way to simplify our choices and cause ourselves the least suffering. In reality, judgments keep us stuck, not at peace.

Mindfulness means adopting a nonjudgmental approach to experiences, just are as they are: they exist; they're neither good nor bad. As you grow into mindfulness, you can catch yourself when you judge your experiences, gently labeling with "judging" when judgmental thoughts go through your mind.

Compassion

A common block to mindfulness is how we approach our own experiences. When happy, we fear the experience will end and so become consumed by greed and the need for more pleasure. When suffering, we fear the experience won't end, and we become consumed by blame, shame, and self-critical thinking. Adopting a nourishing and kind approach to all your experiences will help you to better encode pleasure and lessen suffering.

Cultivating a tender, soft, inner voice that's welcoming and open signals to your deepest self that it's safe to be present in that particular moment. For example, you can tell yourself, *I see you* or *It's okay to have this experience* or *This will pass.*

Curiosity

With curiosity, you're neither looking to actively push away an experience or hold onto it. You're in the role of curious observer. Like watching a storm come over the ocean, you see the sky change, the wind pick up speed, the waves break. You just observe the particulars of what's happening for you in this moment.

EXERCISE: Daily Mindfulness Practice

Approach whatever you're currently doing with mindfulness. First dock the phone so you're not distracted by an incoming call or text. Then use the components of mindfulness to let go and fall into this moment, and the next moment and the next. Here are some examples:

> Breathe in and out: What does your chest feel like on the in-breath, as you expand your chest, and how does it feel on the out-breath?

Brushing your teeth: Feel the brushing sensation, the taste of the toothpaste, the swirl of water rinsing your mouth...

Daily chores: Do one chore at a time; feel the chores physically, step by step...

Cooking: Focus on the different food aromas, the textures as you chop vegetables, the heat from the stove or oven as you stir the pot...

Yoga: Move your body with awareness; observe the sensation of stretching in every part of your body...

Exercise: Feel the strength of your body with each movement, and feel gratitude for what your body provides...

Morning coffee or tea: Notice the heat of the cup in your hands, take in the aroma, observe your body awakening...

Drinking water: Quenching your thirst, imagine the water entering your body, the coldness as it travels down to help you function properly...

Showering: Enjoy the splash of water, the soap foaming, the refreshing physical sensations...

Eating: Slow down your chewing to observe the texture, tastes, and sensations in your mouth...

Walking: Focus on what your foot feels like as it hits the ground and comes up, again and again...

Driving: Notice the grip of the steering wheel in your hands as you take in your surroundings and drive attentively...

Parenting: A you spend time together, nonjudgmentally accept yourself and your children, just as you and they are in this moment...

Spending time with friends or your partner: Nonjudgmentally accept yourself and your loved ones, just as you and they are in this moment...

Commit to bringing mindful intention to one of your day-to-day tasks, even if only for a few minutes, and extend this practice from there.

When practiced regularly, mindfulness will help you grow calmer and less stressed over time.

Mindfulness and the Brain

Each time we practice mindfulness, the brain grows toward calmness, well-being, and focus. In fact, a systematic review and meta-analysis found that at least eight regions of the brain, including areas involved in learning, memory, attention, and emotion regulation, were different in meditators (Fox et al. 2014).

Mindfulness appears to calm the stress and fear area of the brain, the amygdala, and helps us to feel better overall. As we teach ourselves to let go, amygdala activation decreases, and we are less reactive as a result (Kral et al. 2018). Research shows that after eight weeks of mindfulness training, study participants reported experiencing increased mood and well-being, and this improvement was correlated with changes in brain regions specifically associated with mood and arousal (Singleton et al. 2014).

And too, you don't have to be in a meditative state for brain benefits to continue throughout your day. After eight weeks of mindfulness training, a group of researchers assessed participants' brain activity with functional magnetic resonance imaging. These scans showed less activation in the amygdala in reaction to emotional stimuli. This decrease in emotional reactivity held true even when participants were not in a meditative state (Desbordes et al. 2012).

Mindfulness meditation is linked to increased gray matter density not only in areas of the brain associated with emotion regulation but also in areas associated with learning and memory (Hölzel et al. 2011). There's compelling evidence that normative age-related decline in memory and focus as well as corresponding neural decline are less extreme in meditators than in nonmediators. For example, long-term meditators have a less marked decline in gray matter as they age than nonmediators (Luders, Cherbuin, and Kurth 2015). And studies on older adults suggest that mindfulness improves certain neural processes, including those that tend to weaken as we age (Isbel et al. 2020).

Mindfulness has widespread impact on multiple regions of the brain and is likely one of our most potent tools to improve brain fog. Every time you monitor yourself and find yourself zoning out, scrolling, or lost in a spiral of thought, you can just bring yourself back to the present moment with mindfulness. You're training your brain to disengage from the irrelevant and come back to what truly matters.

EXERCISE: Bring Mindfulness to Self

For a true refresh and refocus, the next time you find yourself with a few free minutes, stop and be with yourself in body and mind.

Find your breath.

Feel your chest rise as you inhale.

Feel your chest fall as you exhale.

As your attention wanders, gently redirect your focus to your chest rising and falling.

Stay focused on your breath for a few moments.

Now gently scan your entire body.

Start at the top of your head and go all the way down to your toes.

Nonjudgmentally note the sensations—areas of lightness, heaviness, tension, pressure, tightness, ease, butterflies...hot, cold...

You're not trying to change anything.

Just notice with warmth and acceptance.

Keep focused on your body sensations for a few moments.

Now be aware of the whole you with compassion for whatever you observe.

What does it feel like to just sit with yourself, to be in your body right now?

Instead of robotically turning to social media or searching the internet for a break, use such moments to decompress and remove—rather than compound—your stress.

If practiced regularly, this exercise will help you be more intimate and at ease with yourself. Over time, you'll find a calm center within you that you can return to over and over again.

Putting It All Together

Even if you've been mindless for days, weeks, or years, you can in this moment come back to the here and now. Approaching yourself and your experiences mindfully and accepting your experience just as it is changes the brain's wiring toward calmness, clarity, and focus. Now let's turn to another factor that keeps people stuck in brain fog: self-worth. As you come to accept yourself wholly, while taking steps to grow, internal ease and contentment will flourish.

SOLUTION 8

Overcome Low Self-Worth with Self-Compassion

Running from yourself can induce brain fog. You'll know you're running from yourself if you spend time weighing how to get it right, appear good enough, or dupe people into thinking you're better than you believe yourself to be. If this sounds familiar, you may be living a double life. To your friends, family, partner, or colleagues, you seem to have it together, while deep down you harbor serious self-doubt about your very essence. This duality makes it impossible to feel secure, and so you struggle with perfectionism, indecisiveness, not trusting yourself, fear of failure, fear of abandonment, fear of being alone, or harsh self-criticism....

Hiding and camouflaging whatever it is you reject and avoid about yourself takes energy, making it impossible to feel at ease and present. Instead of confronting your self-worth issues, you run scared. You keep yourself busy, consumed, burned-out...anything but acknowledging yourself and what you fear most.

Low self-worth and brain fog can manifest in different ways. You may push extra hard and expect yourself to be constantly attending to everything humanly possible so that people don't see the real you. Alternatively, you may feel that you can't keep up with your own or others' expectations, so you numb out, going through the motions of life and doing the bare minimum to keep afloat, but without taking actionable steps to improve your situation. A third option is to oscillate between pushing hard and numbing out.

Whatever the case, focusing on your self-worth will improve your brain fog symptoms. This is because overcommitting or numbing out or perhaps doing both means you're overwhelmed and aren't giving yourself what you need to be focused and grounded.

Instead of running away from feelings of low self-worth, you can learn to hold your own hand, warmly guiding yourself through this life. You can cultivate a relationship with yourself where you stay on your own team, even when failing, even when embarrassed, even when defeated, even when alone. As you free up the energy you've squandered hiding from yourself, you'll find a clearer mind and a freer path.

Identifying Your Shadow Self

Most people with low self-worth develop a *shadow self*, a false self that operates to cover up whatever it is they don't accept about themselves, or whatever they despise or disallow about their fundamental nature. This shadow self can take different forms. Here are a few stories showing how the shadow self operates.

Compulsive Business

Sean was professionally successful, always on top of work and encouraging his team members, and passionately supportive of friends and family. He had a reputation as someone who never let anything get him down. In public, Sean was an amazingly productive and engaging person, seen as a rock star by all. In private, he was the complete opposite. When Sean finally went home and had a moment to himself, he'd crumble. He was unhappy, aimless, and filled with self-doubt. He hated being alone because his critical inner voice would mentally replay whatever he'd messed up that day. To avoid himself, Sean filled as many hours as possible with work or with helping others. If he was sufficiently mentally fried, he reasoned, he'd have no mental space to beat himself up later.

Perfectionism

Ever since childhood, Gia had felt like she didn't belong. Her parents had immigrated to the United States when she was in elementary school, and she never quite fit in with her peers. Making matters worse, by the time Gia was in middle school, she was frequently bullied for being overweight. Whether it was her body or her cultural background, Gia's experiences taught her that she stood out, and not in a good way. As an adult, she compensated for these earlier traumas by striving for perfection; if others in her life never saw her flaws, then they couldn't hurt her. She kept her body in svelte physical shape, didn't allow herself to eat outside of a strict diet, and consistently outperformed her colleagues at work. She was exhausted and never at ease within her body or her mind. At the same time, giving up perfectionism was terrifying because Gia believed she'd be "found out" and exposed as the inadequate person she feared herself to be.

Indecisiveness

Jorge was constantly on edge, never able to feel secure and grounded. He was panicked by the fear of making a mistake, so he kept himself in a constant state of limbo. Every time a decision was needed, he'd punt. This included minor choices and demands and also major commitments regarding work and relationships. He'd work himself into a frenzy going over and over a decision in his head, trying to find the "right" or "best" path. Jorge was terrified that if he committed to something he'd miss out on something else. Eventually, decision fatigue would set in, and he'd become stuck, unable to move forward in any direction. Jorge was afraid of making a decision because he didn't trust himself—he didn't trust that if he made a decision that he later regretted, he would be able to pivot, regroup, and still be happy.

Self-Criticism

Helena, a beautiful, intelligent, vibrant young adult, analyzed and scrutinized her every move. She'd be at a beach party with friends, appearing to be happy but in actuality stuck in her head thinking through how she looked in her bathing suit, if any fat was exposed… if she was talking enough…if she was funny enough…how to handle a social issue…how to keep her friends happy with her. This conscientiousness about her social life kept others enamored and eager for her time and attention. However, Helena could never relax or just be herself. Although she had friends, a boyfriend, and plenty of social engagements, Helena always felt alone and unknown. At the same time, Helena feared that if she didn't keep up the relentless self-scrutiny, she'd become invisible to others, not special but just ordinary.

People Pleasing

Lia learned at an early age that to keep her parents' love and affection, she had to be perfectly pleasing. If she had an issue or made them unhappy, they'd withdraw their affection and attention. Over time, she created a false self that was pleasing to them, and by the time she was thirty, pretty much every aspect of her life was about keeping others happy with her. Lia believed that if she didn't consistently put others ahead of herself, no one would stick around. She put so much energy into the needs of others that she had no sense of herself, her preferences, desires, needs, and sufferings, outside of her relationships. She knew this was the case but was terrified of a new way, afraid that she would be left without any love or care.

Lia, like many others, developed a shadow self to avoid looking at her real self or exposing that self to others. Consider for a moment how you operate to avoid whatever you deem unacceptable about yourself. Acknowledge your shadow self and how this way of being takes away reserves needed for focus, internal security, and clarity.

EXERCISE: Show Self-Compassion

Take a few moments to allow whatever is under your shadow self to be present. If you were to give up your shadow self, what would others see about you? What do you fear would be found out and exposed? Invite it in, whatever it is:

Your fear of failure...

Your fear of being seen as a sham, an imposter...

Your fear of others seeing you as average, mediocre...

Your fear of not being seen...

Your fear of not being loved...

Your fear of others leaving you...

Your fear of your internal critic...

Your fear of making a mistake...

Your fear of being alone ...

Your fear of trusting yourself...

How do you feel in your body as you allow your fear to be present? Do you feel tense, jittery? Is your heart beating fast? Do you have an impulse to rid yourself of the feeling? Instead, imagine a warm, loving, compassionate light entering the top of your head and radiating all the way to your toes. This light fills your body with safety and well-being.

Remind yourself that all human beings struggle with negative parts of themselves. There is nothing unworthy or less about you because you struggle with aspects of yourself that you don't like.

Offer warm and loving phrases to compassionately accept yourself. You can say these aloud or to yourself:

"I accept all parts of you, just as you are in this moment."

"I can love you while you experience this upset and fear."

"I see and love all of you."

Acknowledge how hard you've worked to escape your fears and how this has only made you more exhausted and disconnected:

"I get it. You didn't know another way."

"I am here now."

"You're enough just as you are in this moment."

Breathe in and out. Allow these fears about yourself to be here as long as they need to be. You can compassionately embrace this worry, or even hatred, you feel toward yourself. You can observe your own suffering and at the same time offer yourself a soft and intimate presence.

Sitting with your fears may be difficult at first. But as you do this exercise, whatever you are afraid of will start to lose its power.

Your Brain and Self-Worth

Genes, environmental experience, temperament: all interact and create a lasting blueprint for the brain. The ways in which your synapses interact mirror how competent, capable, and lovable you believe yourself to be and have been made to feel over time.

A negative self-image can begin in childhood often as the result of early life stress, trauma, family dynamics, or difficult life events that may make you feel different, unacceptable, or cast out in some way. Also, well-meaning caregivers, coaches, and teachers can make children feel bad about themselves simply by emphasizing outcomes (success, grades, jobs, friends) over process (effort, how you take care of yourself, what makes you happy, how to be alone and enjoy yourself). For others, a series of disappointments in adulthood can lead to low self-worth.

When you chronically feel negatively about yourself, this self-image becomes wired into the brain. Each time you feel vulnerable, hit a setback, or find yourself alone in your thoughts, your brain pulls up this negative self-image and triggers the same old parade of painful thoughts.

Also, the brain carries a *negativity bias,* a tendency to remember and replay negative events in our minds more so than positive ones. This means we could have an extraordinary day when we feel competent and positive about ourselves, but add one negative event into the mix and, poof, we're sucked into dwelling on the negative event and forget all about the positive. This tendency is part of our survival mechanism, as cataloging the negative means we can better anticipate and protect ourselves from threats to our survival—like enormous predators on the prairie. When it comes to our self-image, however, recalling the negative keeps us trapped in negative feelings about ourselves that likely don't even reflect reality.

Feeling not good enough can become such a strong habit that the neurons in your brain develop a hair trigger when it comes to recalling negative stories or feelings of defeat. The neuronal networks in the brain come to predictably fire up the same old pattern of self-doubt and self-criticism. Each new negative thought brings to mind negative thoughts from the past and keeps you in a circular trap.

The brain's negative wiring is maintained because we keep telling ourselves the same negative stories about our worth. We are held in place when we don't give our brains new opportunities to see things differently. Brains adapt and grow only when they work repeatedly at something new.

The process by which you developed low self-worth in the first place is the exact same process you can employ to undo your negative self-image. Activating healthy ways of treating yourself and speaking to yourself starts a new neuronal pattern. Also, each time you engage in new life experiences that make you feel whole, competent, and effective, you start changing the brain's wiring. The more you practice self-compassion and take on novel behavioral actions, the easier it becomes for your brain to cue up positive feelings about yourself.

You can start by meeting each defeat and setback with warmth and kindness, and then take on new activities and initiatives to challenge the negative assumptions you carry about yourself.

Facing Your Internal Critic

To face your internal critic, consider for a moment how you would speak to a beloved child entrusted to your care. If the child hit a setback with friends or schoolwork, would you remind them of every other setback that had ever occurred to them? ("You always do this! I can't believe you messed up again!") Would you yell and shame them when they're upset? ("You're so dramatic and needy...stop!") Would you tell them they're incapable? ("You'll fail at that too...there's no point in trying.") Would you tell them if they don't make sure others are happy with them, and if they aren't perfect, no one will love or want to be with them? Would you encourage them to sit alone in a dark room, eat junk food, and think about how bad their life is? Would you discourage them from talking with others about what upsets them, because no one cares and it's embarrassing?

You're probably thinking, *No way! I would never treat someone like that.* But I ask you to consider now how you speak to yourself. What is your own internal narrative, that little voice inside your head that comments, assesses, and judges pretty much all of your experiences and yourself?

We all have an internal narrative. Like the air we breathe, the internal commentator is always there. And how you speak to yourself, and how you can choose to speak to yourself, has the potential to lift you high or cripple you. If the tone you take with your own self is critical, punitive, shaming, unaccepting, or harsh, then you're not on your own team and you're setting the stage for your own defeat.

Sometimes I ask clients to say out loud what they're saying to themselves about a setback or hardship or difficult feelings they're experiencing. They're often stunned to hear how unforgivingly they speak to themselves. They've existed with a negative narrative for so long that it plays out on repeat, day in and day out, and they've lost sight of how much that harsh internal tone holds them back.

You are your most potent ally. If you're not firmly playing on your own team, it will be hard to pull yourself out of any rut, including brain fog. Maybe the exercises in this book make sense to you, but if your internal voice is beating you up in the process—telling you,

What's the point? This isn't going to work or *I'll never be good enough* or *I have too much to fix about myself*—then you won't have the will to persist in your own growth.

You can change this pattern by cultivating a warm and accepting narrative that unconditionally loves and cares for *you,* even when you're experiencing things about yourself or your life that you don't like. Here's a tool to start understanding how to emotionally support your innermost self.

EXERCISE: Change the Narrative

Imagine someone who truly cares about you—a parent, teacher, friend, grandparent, sibling, therapist, romantic partner—and what they might say to you in your challenging moments. Say this to yourself now.

Or call to mind a time in your childhood when you really felt awful about yourself. Can you remember feeling left out or not good enough? Now what do you think you needed to hear from yourself or from someone who cared? What would you have liked to believe deep within your core that would have helped you get through that adversity while still valuing yourself? Say that exact thing to yourself now.

Here are some internal comments you can use to stay on the same team as your innermost self:

You're an amazing human, just the way you are.

You are enough.

We will figure this out.

Even when you struggle, you have worth.

You're not alone; others have felt just the way you do right now.

I love you.

I'm always here and will help you find the way.

Perfect is boring. Be yourself and you will see all the good that comes.

Everyone has hardships and difficulties, so you're not alone.

You're worth everything good in the world.

Your value is so much more than what people think of you.

You can grow and improve on the things that you struggle with.

I love all of your quirks; they make you interesting.

You belong.

You are stronger than you know.

Struggling with things makes us human.

Talk to someone about what you're feeling, and you'll feel better.

Ask for help with this. It's too much to manage on your own.

Each time you become aware of a negative narrative playing out, call to mind a softer, gentler tone and phrase. Building awareness and adding one simple phrase starts a new pattern, a new way of relating to your self. Comforting and supportive language will eventually cue up automatically.

Taking on New Experiences

Self-worth is not only contingent on changing the story you tell yourself in your own head but also on the behavioral actions you take on a day-to-day basis. When your experiences of yourself remain the same, it's easy to use this as evidence of your low worth. In reality, your experiences of yourself are the same because you're doing the same things and are around the same types of people. It's important to both accept yourself wholly and work on specific areas of growth.

You can start this process by listing specific growth points you want to work on and the positive steps you can take to increase your self-worth. Here are some suggestions to get you started:

Confronting compulsive business. Take one evening or afternoon a week to yourself, and find ways to enjoy yourself, all on your own. Journal, eat an enjoyable meal, watch some TV, take yourself on an outing…whatever it is, learn to healthfully be with yourself.

Opposing perfectionism. Take on a new endeavor without the intention of being noticed, doing well, or feeling praised. This is simply for your own pleasure.

Challenging people pleasing. Carve out a space to start considering the larger goals, desires, and needs you have outside of making sure others are happy with you. Resist the urge to compulsively sacrifice yourself for someone else. Instead, ask yourself, what do I need in this moment?

Escaping work ruts. Talk to your employer, a mentor, or job coach about what would help you feel more energized at work, or consider applying for a new job or embarking on training or an academic degree that will help you grow professionally.

Handling helplessness. If you feel incompetent and defeated, take on a new challenge. Volunteer at your kids' school or for a community initiative, tackle a DIY project for your home, or organize a closet to clear out some room. Run for a local political office or start a task force for something that's meaningful to you.

Getting out of social ruts. Join a social group, exercise group, book group, or something that exposes you to different types of people who will see you, and help you to see you, in a new light. Making new friends is tremendously valuable for increasing self-worth.

Encountering indecisiveness. Commit to something small—a garden, a pet, an exercise routine, a friendship, a volunteer initiative, just for yourself—no one else needs to know about it. Stay committed even when you don't feel like it.

Cutting back on self-criticism. Get out of your comfort zone. Try something new. Pick something you've avoided in the past because you've been afraid of what others would think.

Working on romantic relationships. Consider therapy or reading a self-help book on improving communication or intimacy. Or consider whether you need to end a relationship that is holding you back and only reinforcing your sense of worthlessness.

Putting It All Together

No one feels perfect all of the time. The best you can hope for is a safe and comfortable house for your innermost self to rest within. This means accepting yourself as you are and knowing that with effort and new experience, you can change long-held negative self-perceptions. As you learn to relate to yourself with greater self-compassion, you may become aware of the deficit of fun and playfulness in your life. Next we will explore how to increase spontaneity and joy.

SOLUTION 9

Overcome Lack of Joy and Pleasure with Creativity and Play

I often hear clients prioritizing work and career over leisure and fun. Some are committed to getting ahead and making enough money while young so they can retire early ("Then I can finally relax and have some fun"). Others put off having fun until their youngest gets to college ("Then I'll have the time to start my ceramics hobby"). Yet others motivate themselves to work endless hours with the reward of an eventual vacation ("Well, in January I can unplug"). Does this sound like you too?

This kind of separation between work and play is a fantasy that rarely delivers. If you can't relax and have fun in your day-to-day life, it's quite difficult once on vacation to suddenly let loose, or once in retirement, to know yourself outside of work well enough to build a happy life. Instead, you end up on vacation with your brain still in high gear: *The ocean's too cold, and where is the server?!* Or, you end up finally retired, thinking *I have absolutely no idea what to do with myself without my job.* Or your kids finally leave the nest, and you find yourself alone with your partner with no sense of who they are or how to make each other happy. The other problem with this approach to work and play is that, until the vacation, until the retirement, until the kids leave, you feel dulled out, old, and deenergized. You may carry the nagging sense that somewhere along the way, you lost the point of life. Working hard all the time may leave you at a loss for coping with day-to-day stressors.

If you don't make time for play, then work ennui, lack of motivation and inspiration, and feelings of dissatisfaction will inevitably set in. You'll go through the motions of life and subsist on those fantasies that all the work will eventually lead to a pot of gold at the end of the rainbow. In actuality, the days turn into weeks, weeks into months, and months into years of all work and no play. Playing *now* pays off in the present and in the future. Making room for play and creativity increases psychological well-being and intimacy in our relationships, and it makes us happier, improves brain functioning, and quite simply makes life more worth living.

Recognizing the Value of True Play

The very idea that the modern-day reward for managing our various tasks and responsibilities is that we get to zone out on our phones or have a glass of wine later is an abject neglect of the human spirit. Phones, tablets, computers, and video games are all easy to access but also devoid of meaning, empty of novelty, and draining. Instead of refreshing our spirits, we end up with screen fatigue, which like TV bingeing, overeating, and alcohol and drugs reduces focus and concentration. Once we return to the task at hand, we're even more tired and distractible, and it takes even longer to accomplish our goals.

Unfortunately, zoning out is all too common these days, because when healthy outlets for pleasure and joy are absent, we gravitate toward these self-defeating means. When was the last time you really played?

EXERCISE: Recall the Little You

To get in touch with what brings you lightness, fun, and pleasure, take a moment to consider yourself as a child. Can you recall what you enjoyed doing as a kid? Did you like being outside, spending time with animals, making art, playing games, using your imagination? Do you recall how you felt when you were in the play zone as a kid?

When is the last time you had this feeling as an adult? What could bring about a similar feeling in your life now? Can you imagine scenarios where you might feel some lightness, fun, and pleasure again?

It's easy to see the pleasure that play and creativity bring about in children. Take a grumpy kid and start to play with or make art with them, and you'll see the magic. That moments-ago grumpy child is happy once again, optimistic about the day, energetic, and connected to you. What's true for the child is also true for the adult. Even with a few minutes of laughter or silliness, we emotionally recenter, feel hopeful, and become more positive.

Playfulness, pleasure, and creativity increase our zest for life, psychological well-being, resiliency, and ability to cope with stressful life events. Adult playfulness is strongly linked with life satisfaction, including reported physical and psychological well-being, pursuing pleasurable activities, and generally engaging in an active life (Proyer 2013). Playfulness also helps us put stressful events into perspective and cope with life's ups and downs. For example, playful adults experience lower stress levels than others and are more likely to have adaptive, positive coping styles. While less playful adults are more likely to blame themselves for their stress and avoid or do nothing to improve it, playful adults are more likely to deal with stress directly (Magnuson and Barnett 2013).

Resilient people tend to naturally use positive emotions to better manage their response to stress and to improve their mood state. This makes sense in light of research showing that even the anticipation of a pleasurable event can help you bounce back more quickly from stress. For example, if you know you're going to have a nerve-racking day but you also have a plan for some fun at the end of the day, you'll likely cope better and experience less stress as a result. Anticipating a future event that will elicit positive emotion is a powerful way to decrease the body's stress response.

Experiencing joy and pleasure reenergizes and reinvigorates our deepest self. And, feeling as if we have outlets for our own contentment keeps us interesting and more alive. Without creativity, fun, and playfulness, adults become dull. Incorporating pleasure and playfulness into your life will make you more easygoing and fun to be around.

A Happy Mind Is a Focused Mind

Psychologist and researcher Peter Gray (2014) describes in his TEDx talk "The Decline of Play" how juvenile mammals of all species conduct a sort of dress rehearsal for adulthood through playing. Gray talks about how play is a way for these animals to condition their bodies and regulate their emotional responses and learn how to cooperate, along with other social skills. When researchers deprive juvenile mammals of play, the animals become markedly different; they can't effectively manage the social signals of the other animals, and they struggle with regulating their fear response.

In humans, just as with other mammals, childhood play is essential for brain growth, neuronal connections, and developing the emotional, cognitive, and social skills needed in adulthood. In fact, as brain size correlates with the need for play, human children play far more than any other mammal (Gray 2014). From an early age, children naturally play and gravitate toward learning about their environment through play.

Furthermore, the need for play and healthy brain functioning doesn't stop in childhood. Studies show that cognitively stimulating leisure activities, including number games or crossword puzzles, delay or diminish cognitive decline and may reduce the risk of dementia later in life (Litwin, Schwartz, and Damri 2017; Yates et al. 2016). Also play and pleasure release feel-good chemicals in the brain, including endorphins, that decrease stress and increase a sense of well-being. The simple act of laughing triggers the chemical changes in the brain that decrease anxiety and increase tolerance for pain and stress (Manninen et al. 2017; Louie, Brook, and Frates 2016).

Playing, creating, being silly, laughing—all are momentary ways to let go of your adult life. Even viewing a few humorous social media memes increases positive emotion, decreases stress, and increases confidence for managing difficulties (Myrick, Nabi, and Eng 2021).

If you think about it for a moment, you've likely had the experience of working hard on something—perhaps struggling to find a solution to a difficult problem or finding yourself unable to start a dreaded task or just feeling exhausted from your responsibilities—and for one reason or another, you get pulled into something silly or playful or you find yourself giggling with a colleague or friend. Then, upon returning to your tasks, they no longer seem quite as intense. Suddenly you're able to see new possibilities, solve a particular issue, or get yourself to focus and concentrate.

When we're pulled out of ourselves and into the joy and whimsy of the present moment, the brain resets. This was shown when a group of Harvard Researchers tracked happiness through the "Track Your Happiness" app. The app pings participants throughout the day and asks them to rate their happiness with what they're doing in that specific moment and report whether their mind is distracted or focused in that moment. These researchers found that participants spend approximately 47 percent of their time thinking about things other than what they're doing in the moment. And, most importantly, this "mental mind wandering" is a significant predictor of a person's unhappiness (Killingsworth and Gilbert 2010). According to this research, whatever the activity, we're happier if our brain is focused on what we're doing than if our mind has drifted elsewhere. And, of course, pleasurable events bring about a stronger focus on the present moment. One clear example of this is that the happiest people in this study were the ones who were having sex at the moment their app pinged them for a response!

When present, we're happy. This phenomenon is also seen in *flow*, a state of being where we experience ourselves as fully present, utterly alive, and completely in tune within a task (Gold and Ciorciari 2020). In a flow state, attention, emotion, and focus are all in line entirely without self-consciousness. People get into this kind of zone with all sorts of activities: music, athletics, sex, religion, and creative endeavors including art.

Of course, some behaviors, like sex and flow, aren't possible all the time! But this research is a reminder that engaging in what we enjoy counters the negativity bias, or, as explored in chapter 8, the brain's tendency to remember and replay negative events more so than positive ones. When present and focused, we aren't replaying events from the past or worrying about the future.

Making a deliberate effort to encode the moments of joy and pleasure in your life, to make room for spontaneous play and creativity, and to recall and anticipate positive events rewires the stressed-out brain. Engaging the fun around you is protective and will take you out of your stressed-out existence. So...how do you get this giant brain refresh? Let's turn to that now.

Remembering How to Play

Letting go of all that adulting, even for brief periods of time, is what play is all about. This idea is perfectly epitomized in the Japanese concept of *gachapon,* toys designed for adults (Dooley and Ueno 2021). Available through Japanese vending machines, a random spin of the dial and, ta-da, a tiny toy object appears! The objects themselves are of the strangest ilk because they are so *not* strange but entirely ordinary and commonplace. Gachapon items include tiny gas cans with functioning nozzles, miniscule air conditioners complete with ducts and fans, and mini shaved-ice machines, to name a few. These objects are funny because of the unexpectedness of which object the vending machine will deliver, and there is also humor in seeing an everyday, random object immortalized in miniature. Gachapon toys bring a sense of levity and silliness to the Japanese consumer.

Play and creativity are present in truly any activity in which a sense of lightness, pleasure, humor, lack of self-consciousness, or novelty permeates. And most importantly, in play mode, we are not goal or result oriented but instead firmly committed to the intrinsic joy of the experience itself. There's no purpose other than the pleasure of the particular moment.

So, play is not goal-oriented: doing ceramics to appear creative to your friends or running to win a marathon or cooking to put a meal on the table. These are each fine goals, but play is doing ceramics because you like the feel of the clay in your hands. It's running to feel the strength of your body and your physical self. It's dancing while you cook, because your brain is in the moment and happy.

Here are a few more things you can do to bring more play into your life:

Take a bike ride

Joke around

Play tag

Look at art

Create art

Color in an adult coloring book

Doodle

Watch a comedy skit

Listen to music

Sing

Play a musical instrument

Dance

Explore a new place

Woodwork

Read poetry

Collect something

Write just for yourself

List good things about yourself

List good experiences from your past

Have sex

Do crossword puzzles, board puzzles, mind games

Be silly

Gigi remembered her love of arts and crafts and took up painting again. Simon recalled horseback riding and started going to a local barn where he could ride. Cecile recalled her love of nature and started caring for bonsai trees. Jack remembered being silly and funny as a kid and started doing stand-up comedy at a local open mic event. Madeline loved running around her neighborhood as a child, so she took up ultimate Frisbee. Josh recalled his love of Legos and started restoring old boats.

What will you choose? There's no one right answer. The key is to connect with feelings of joy and lightness from your past and figure out how to bring more of that into your present life.

Adopting a Playful State of Mind

One reason we don't allow ourselves the space to be free and silly and creative more often is that as adults we're conditioned to see playful leisure as unproductive, a waste of time, or even meaningless. Also, many adults feel selfish taking time for themselves that isn't goal oriented or in the service of their family, work, school, or other responsibilities. If you feel this way, then it's important to change your mindset. Research shows that perceiving free time as wasteful impairs your ability to actually be present and to enjoy free time and is associated with a decrease in psychological well-being, including increased anxiety and stress (Tonietto et al. 2021). Believing that free time is productive and helpful to you is important, so you can really encode the experience.

Playful activity and creativity don't deliver an end goal as tangible as completing a work project or cleaning out your basement, but engaging in something that's not result-driven paradoxically increases our efficiency and persistence when we do return to our goals. We're more productive and better able to see solutions and understand what needs to be done. In addition, when we add some levity to our life, we have more vitality for our day-to-day demands. Also play is a way to connect with loved ones. Every now and again in the most random of moments, I throw my nine-year-old daughter onto the bed and make a life-size burrito out of her with pillows and blankets. The silliness of this unexpected action causes us both to erupt into uproarious laughter. When we let go and have fun together, we feel close and connected.

You can adopt a positive mindset about pleasure and play: it is not a waste of time but rather a way to improve your life, your psychological well-being, your stress, your relationships, and, yes, your brain. Each time we let loose in a belly laugh, engage in silliness, or get creative, we clean the slate of the brain; we are refreshed and ready to effectively engage.

EXERCISE: Consider Your Work-to-Play Ratio

When the tasks, responsibilities, and goals of life outweigh the pleasure and rewards, brain fog sets in. You start to feel like, *What's the point of all of these accolades, all of this work, all of this money, this new house or job promotion, when I feel sad, dulled out, and disconnected?* Consider the ratio of work to pleasure in your life. If you picked up this book, chances are it's out of whack. There needs to be a balance of the stressors—deadlines, responsibilities, laundry, family commitments, bills—and the perks, like enjoying the moment, relaxing, laughing, passion, fun, friendships, creativity. Ideally, the emphasis should tilt toward the latter. When you have enough fun and novelty in your life, then it's easy for the brain to anticipate positive events, and this buffers the negativity bias.

Take a moment now and consider how you enjoy life. Is there room for more joy? What stops you from making in-the-moment fun a priority? Are you living for some eventual pot at the end of the rainbow that may never materialize? Consider the idea that you don't have to wait for a vacation or retirement. You can live your best life right now and likely increase your ability to focus, concentrate, and problem solve. And, you will be happier and more fun to be around as a result.

It's easy to get caught up in the negative. But if you are open to something different, there are countless opportunities for play and creativity. The key to rewiring the brain for pleasure is to encode these experiences while they're occurring. When we acknowledge our pleasurable experiences—when we name them—we're giving our brain a moment to catch up with how good it feels to laugh, to play, to be spontaneous, to be present in an intrinsically pleasurable task. So the next time you find yourself in a happy or a peaceful or a laughing moment, pause to give your brain a chance to take it in. Acknowledge the experience: *It feels good to laugh* or *I am present and happy.* Bring your full self into the moment and note how good it feels so that you can rewire your brain toward joy.

Putting It All Together

The last tool for coming out of the fog and fully into your life is to make room for joy and pleasure. Adult life is often serious and goal driven—everything is to get somewhere else or to get something else—but when we smile and laugh and relax, once again we feel young and alive. You can experience this giant clearing of the soul and start anew, not just after a stellar vacation but in your day-to-day, moment-to-moment experience of yourself.

Let's turn now to putting together all of the tools you've learned so far so you can take the steps needed for coming out of the fog and into the clear light of today.

SOLUTION 10

Overcome Brain Fog with Actionable Steps

With your new tools handy, you're ready to put brain fog in the rear-view mirror. You're in the driver's seat again. You have likely clarified your intention, understand what needs to be done, and recognize that implementing the strategies in this book will work. But...are you fully motivated to start taking action and persistent enough to carry out the steps you need to take? Hopefully, you're feeling inspired, but it would also be completely normal if you, like many, have been struggling with the day-to-day discipline needed to make these solutions habits.

For a long while now, your life has been too full and too heavy to find a place for new behaviors that would reduce your level of stress. Changing the way you do things may have felt like a waste of time, unnecessary, when so much else was overwhelming and demanded immediate attention. It's also possible that you never really had a road map for healthfully managing stress. Perhaps you got away with one way of being for a long time. Either way, brain fog sets in when what you're asking of yourself—or what those around you or the environment as a whole is asking of you—exceeds your internal resources. Denying yourself the emotional nourishment you need has consequences: forgetfulness, anxiety, distractibility, sluggishness, lack of organization, and irritability, to name a few.

As we've explored, when chronically stressed you may live out a life on perpetual repeat, where even though you may know that the way you're spending your time isn't healthy, you wake up and do the same thing over and over again. The neuronal mapping leads the way, and before you know it engaging in unhealthy routines becomes your brain's default position.

My hope for you is that you now have a road map for how to rebuild or build for the first time a strong bank of internal resources. As you develop these resources, you'll find yourself heartily ready for the adversity, conflicts, responsibilities, and upsets as well as the wonderful pleasures and rewards that life can deliver. This solution will explore the steps you need to take to bring these brain fog solutions to life and into your day-to-day routine; it will also cover how to maintain your new commitment to yourself, even when you want to give up.

No matter what you've experienced in the past, no matter what you regret, no matter what you feel poorly about, you have the capacity to start again, to make a change, to create a new beginning...right now. Are you open to a fresh start? Let's begin!

Committing to a New Path

At times, you've likely felt overwhelmed as you've made your way through this book. You wonder how you'll fit all of this self-care into your life. You may doubt whether you have the time and even ask yourself if it will really pay off. The answer to this is to hold in mind that what we do behaviorally, whom we connect with, and what we tell ourselves impacts how we feel and how we think. If you agree with this, then it's a far less overwhelming decision to just spend more time than you have previously on your own self-nourishment.

If you make that choice, if each day you have on your agenda even one or two of the exercises in this book, you'll see the results. Each time you engage in a new way of thinking—relating to yourself, taking care of yourself, connecting with others—you're turning your brain in a new direction. Little by little, day by day, as you make that choice to invest in yourself, you're changing your neuronal wiring. Over time, your deliberate actions will become habits that automatically cue up, steering you toward resiliency and well-being.

Check in with yourself and see if you can internally commit to the idea that every time you take an actionable step, no matter how small, you're on the path that leads away from brain fog. Make a silent internal commitment to yourself that you're going to deliberately incorporate the solutions into your day-to-day routine and that if you forget to do this—if you momentarily lose sight of the goal or want to give up—you will recommit again and again.

Confronting Blocks to Growth

Even with the best of intentions, we humans have all kinds of limitations that get in the way of developing new habits. As you reflect on what it means to make this kind of commitment to yourself, ask yourself what blocks might get in the way of decreasing the stress in your life. What blocks might get in the way of you reacting differently to the stress that you can't control? What might sabotage you from taking on new behaviors? Try also to imagine the ways in which you can circumvent these blocks to making change. For example, if lack of time is a problem, could you wake up a half hour earlier? Or, if lack of accountability is a problem, could you tell a friend and ask them to encourage you?

Let's put a closer spotlight on some common blocks to growth—including self-judgment, waning motivation, not knowing where to start, lack of accountability, and lack of persistence—and how to move beyond them.

Dealing with Self-Judgment

There will be days when you forget entirely about your work to escape brain fog, there will be days when you're too exhausted to think of yourself at all, and there will be days when you simply don't want to do what a huge part of you knows needs to be done. And you may very well experience certain pressures, including economic, professional, and social adversity, that make self-growth harder to implement or even impossible.

First, consider your expectations for yourself and whether they're reasonable given the parameters of your life. It's common for people to beat themselves up for not being more perfect in their pursuit of growth and to underestimate the work and time true change takes. Try not to judge yourself if you decide that making this kind of change is too much to do right now. Likewise, if you've decided to commit to change because you believe you can change, do not judge yourself harshly if you don't always live up to your daily goals.

Unreasonable expectations typically manifest in the form of self-judgments and internal criticisms: *Ugh I did it again. I suck. I can't do anything right. Something's wrong with me. I am permanently screwed. I am a loser. I read that whole book, and I've done nothing.* It's precisely these self-judgments that cause people to self-sabotage, give up, and go back to their old ways of thinking and doing.

No one is perfect, and no one perfectly changes long-term behavioral habits. Success in your growth doesn't mean you will face no setbacks. Instead of judging yourself in response to inevitable challenges, act the part of a skilled basketball player who misses an important shot. Let it go so you can successfully stay in the game and focus on the next shot.

Move on from the past and into the present; simply recenter and start again. Warmly embrace the future and the future choice you have to reinvest in yourself. When outside factors are simply too powerful to go against, give yourself compassion and acceptance. When the window opens again and you can get a breath of fresh air, you'll start again.

What to Do About Waning Motivation

On your path to clarity and peace, your motivation will wane at times. You may even forget why you read this book or what it was you were after in the first place. Getting a sense of where you are right now, and even writing it down, will help you later if ever you find your motivation is waning. First consider these two scenarios.

James's Story

James was completely burned out. He worked morning, noon, and night. He logged in for work at 6:00 a.m. most days and often stayed in front of his computer for eight hours at a time. He rarely stopped to walk around, have a snack, or even use the bathroom. He'd take a break to eat dinner with his family, but then he'd get back online, often working until 1:00 a.m. He felt constantly on edge. Even while sleeping, he'd wake in the middle of the night feeling physically keyed up. He was plagued by uncertainty and had trouble making decisions. He was so mentally exhausted that he never had time for fun or for the people he cared about in his life. When he did have a free moment, he would typically zone out in front of the TV or on his phone. Then he'd beat himself up for wasting time and feel guilty and lazy for not being more productive. James would remotivate himself by remembering that if he worked hard enough, he could retire early and finally be able to relax.

Aidan's Story

Aidan was very focused while at work but had a hard cutoff at 6:00 p.m. When he ended his workday, he knew there was more he could accomplish, but he wanted his life to be more than just work. After work, he'd change his clothes and go into relaxation mode. Aidan would help with his kids' homework, play games with his family, or enjoy watching a movie with his partner. As full as life was, Aidan spent some time every day on something for himself. This could be as simple as meditating for ten minutes, doing some light exercise, or checking in with himself physically and emotionally. On weekends, he rarely worked. Instead, he enjoyed going to boat shows and spent time refinishing old boats. Aidan and his partner were close and liked spending time *together as well as with their kids and their friends. Aidan was content and grateful. He occasionally thought about retirement but generally wanted his current life to go on as long as possible.*

How does your story compare with these? Do you relate to either? Take a moment now to reflect on your own story, where you've been and what you're turning toward, so that you can remind yourself when you need it most.

EXERCISE: Writing Your Story

Write in your journal your story *before you read this book*. Call to mind how you felt. Were you agitated, stressed? What did you feel like in your physical body—tense, unhealthy, agitated? Consider how mentally exhausted you might have been. How did you feel emotionally—stressed, forgetful, disorganized? How do you think your loved ones perceived you? How present were you in your life and in your own body?

Now write your *new story going forward*. How do you want your outside life to look, and how do you want to feel on the inside? Consider your physical self and what you're turning toward. Is it physical ease and calmness? Describe whatever thoughts or images come to mind. What do you want your emotional self to radiate? Peace, well-being, presence? Use your own words to describe how you imagine yourself feeling emotionally. How would your life look if you were more at ease? What kinds of things would you be able to do that you can't do now?

Coming back to your new story, again and again, will help inspire your path going forward and your new way of being.

Figuring Out Where to Start

You've made it this far, so you're likely convinced that employing these tools will be good for you. At the same time, it's overwhelming to think of where to start. One way is to create a master list of tools that will help you move away from mental exhaustion and toward peace and clarity.

TOOLS FOR OVERCOMING BRAIN FOG

Here is a list of many of the tools we've discussed in this book:

Make a list of what's in your control and not in your control.

Make a list of actionable steps you can take to improve a problem you're facing.

Accept uncertainty as an inevitable part of life.

Take fifteen minutes to focus on larger goals you have for yourself.

Call a friend and focus completely on that conversation.

Talk to a friend about the difficulties you're facing.

Volunteer.

Walk.

Reduce screen time.

Don't look at your phone first thing in the morning.

Reduce alcohol consumption and recreational drug use.

Reduce bingeing on TV, video games, or pornography.

Invest in a social pursuit.

Turn off technology and read a book before bed.

Share something with a friend that you wouldn't ordinarily share.

Take the time to make small talk.

Imagine a warm, compassionate light entering the top of your head and radiating all the way down to your toes.

Take ten minutes for an emotional check-in with yourself.

Express your feelings through talking or writing.

Inhale slowly for a count of four; exhale slowly for a count of four.

Call to mind a calming visual image as you breathe in and out.

Do progressive muscle relaxation: clinch each muscle in your body as you inhale for a count of three; release each muscle as you exhale for a count of three.

Focus on your nutrition.

Focus on healthy sleep.

Exercise.

Do an activity that relaxes you.

Spend fifteen minutes in nature: observe the leaves, the trees, the sky, the clouds, as you breathe in and out.

Reduce or eliminate caffeine.

Observe your thoughts.

Challenge unrealistic thinking.

Notice when you're catastrophizing.

Recognize faulty conclusions.

Give up rumination in favor of active problem solving.

Journal your thoughts.

Mindfully eat.

Focus on each of your senses.

Practice acceptance.

Mindfully be with yourself for ten minutes, focusing on your breath and your body sensations.

Offer yourself warm and loving phrases.

Take on a new experience.

Introduce fun back into your life.

Remember what it was like to have fun as a kid.

Work to increase pleasure in your life.

List or make a mental note of all the good in your life.

Stay on your own team.

If some of the solutions for brain fog in this book appealed to you more than others, then put them on your master tools list. Once you know what you want to focus on, you can create a weekly schedule to put your ideas into action.

EXERCISE: Create a Weekly Schedule

Look at the coming week and see if you can put into your calendar one or two new actions that you're willing to take each day. If you forget or don't take the action, just pick up and start again the following day. Most of the exercises we've explored take fifteen minutes or less. Physical exercise is often a bigger commitment, but remember that even fifteen minutes of vigorous exercise is good for your brain.

Weekly schedules can be quite simple and to the point. Here's an example of one:

Monday: *Ten minutes of mindfulness meditation. Drink a smoothie loaded with fruits and veggies.*

Tuesday: *Take a fifteen-minute walk around the block. Check in emotionally with self.*

Wednesday: *Call an old friend; be present and share. Watch a comedy skit.*

Thursday: *Record thoughts and identify any faulty conclusions. Place sticky notes in key places with compassionate and accepting thoughts.*

Friday: *Take a warm bath and mindfully relax. Take fifteen minutes to write down larger goals for work, family, or social life.*

Saturday: *Read a good book. Mindfully eat each meal today.*

Sunday: *Go for a mindful hike in nature. Volunteer at the local food bank.*

The simple act of making a space for yourself, even if only fifteen minutes a day, sets your intention to be grounded in your body and present in your life.

Sharing with Others to Stay Accountable

One reason brain fog endures is because people feel they have to hide it. This is when imposter syndrome comes in. You pretend to your colleagues, your family, and your close friends that you have it together more than you really do. Keeping your brain fog a secret saps you of energy, whereas sharing with others brings in new energy, fueling your psyche to keep you motivated and on track.

Whenever we invest in a new habit, we are far more successful when we allow others in on what our goal is for this new habit. The simple act of saying it aloud makes it more real and will make it harder for you to avoid and deny what you want for yourself.

Create a true support system by sharing with others that you've read this book and that you recognize your brain is foggy. Also share some of the goals you have for improving your thinking and energy and what you're working on now. Make a point of trying to hold yourself accountable by bringing up the conversation and letting a few folks in your life know of your progress, your setbacks, what you're discovering about yourself, and what you need to be successful in your journey toward clarity.

Setting up reminders on your digital calendar or putting sticky notes in strategic places, like your mirror or car dashboard, will also help keep you on track. The simple note to "breathe" is a cue for the brain to remember to make a space for yourself.

Review your master list on a monthly basis. Make a mental note of what you've included in your routine and what you're avoiding. Then renew your commitment to yourself and start again.

Sticking with It

As we've explored, our brains aren't thrilled when asked to do something differently. It takes effort to cue up new patterns and ways of being. This is why growth is bumpy and never as easy as we think it should be. This is also why we become tired and lazy and want to go back to our old easy patterns.

When you want to give up, remind yourself that growth is not possible without feeling uncomfortable. In fact, feeling uncomfortable is a sign that you're growing, that you're pushing yourself out of your comfort zone, and it's exactly when we push ourselves out of our comfort zones that we change.

Success in your quest for peace, calm, focus, and presence is dependent on you believing, deep within your core, that it will pay off if you keep playing the long game. And even when you have setbacks, bad days, or feel awful in spite of working hard, it doesn't mean you're failing. Instead of looking at your progress on a day-to-day, moment-to-moment level, take inventory a month from now, two months…six months. In this way, you can focus on the bigger picture of your growth and not get stuck in the details of what you aren't doing well.

And most importantly, when you hit a setback or forget your goal of coming out of the fog, simply recenter and begin again. In time you'll see a clearer and brighter pattern emerge.

Putting It All Together

In this chapter, you made a commitment to yourself to start anew. You know that you, today, at this moment, have the power to do this. In addition to understanding the solutions to brain fog that we've explored in this book, you now have a concrete plan for how you're going to bring these solutions into your day-to-day life. You don't need to devote hours or days to the task or be perfect in your pursuit of growth. But you do need to be committed. When you want to give up, give yourself warmth, compassion, and acceptance. Then, allow yourself to recommit and get back on the path to peace, again and again.

REFERENCES

American Sleep Association. 2022. "Sleep and Sleep Disorder Statistics." www.sleepassociation.org/about-sleep/sleep-statistics/.

Bavelier, D., M. Dye, and P. C. Hauser. 2006. "Do Deaf Individuals See Better?" *Trends in Cognitive Sciences* 10 (11): 512–18.

Baycrest. 2017. "Baycrest Creates First Canadian Brain Health Food Guide for Adults." https://www.baycrest.org/Baycrest-Pages/News -Media/News/Research/Baycrest-creates-first-Canadian-Brain -Health-Food.

Davis, D. M., and J. A. Hayes. 2011. "What Are the Benefits of Mindfulness? A Practice Review of Psychotherapy-Related Research." *Psychotherapy* 48 (2): 198–208.

Desbordes, G., L. T. Negi, T. W. W. Pace, B. A. Wallace, C. L. Raison, and E. L. Schwartz. 2012. "Effects of Mindful-Attention and Compassion Meditation Training on Amygdala Response to Emotional Stimuli in an Ordinary, Non-Meditative State." *Frontiers in Human Neuroscience* 6: 292. https://doi.org/10.3389/fnhum.2012.00292.

Dewall, C. N., G. Macdonald, G. D. Webster, C. L. Masten, R. F. Baumeister, C. Powell, D. Combs, D. R. Schurtz,

T. F. Stillman, D. M. Tice, and N. I. Eisenberger 2010. "Acetaminophen Reduces Social Pain: Behavioral and Neural Evidence." *Psychological Science* 21 (7): 931–37.

Dooley, B., and H. Ueno. 2021. "A Tiny Gas Meter? The More Mundane the Better for Japan's Capsule Toys." *The New York Times.* October 8. www.nytimes.com/2021/10/08/business /japan-capsule-toys-gachapon.html.

Fox, K. C., S. Nijeboer, M. L. Dixon, J. L., Floman, M. Ellamil, S. P. Rumak, P. Sedlmeier, and K. Christoff. 2014. "Is Meditation Associated with Altered Brain Structure? A Systematic Review and Meta-Analysis of Morphometric Neuroimaging in Meditation Practitioners." *Neuroscience and Biobehavioral Reviews* 43: 48–73.

Gold, J., and J. Ciorciari. 2020. "A Review on the Role of the Neuroscience of Flow States in the Modern World." *Behavioral Sciences (Basel, Switzerland)* 10 (9): 137.

Gray, P. 2014. "The Decline of Play and Rise of Mental Disorders." *TEDx Talks,* June 13. https://www.youtube.com/watch?v=Bg -GEzM7iTk&t=135s.

Harvard Study of Adult Development. 2015. https://www.adult developmentstudy.org.

Hölzel, B. K., J. Carmody, M. Vangel, C. Congleton, S. M. Yerramsetti, T. Gard, and S. W. Lazar. 2011. "Mindfulness Practice Leads to Increases in Regional Brain Gray Matter Density." *Psychiatry Research* 191 (1): 36–43.

Isbel, B., J. Weber, J. Lagopoulos, K. Stefanidis, H. Anderson, and M. J. Summers. 2020. "Neural Changes in Early Visual Processing After Six Months of Mindfulness Training in Older Adults." *Scientific Reports* 10: 21163.

Killingsworth, M. A., and D. T. Gilbert. 2010. "A Wandering Mind Is an Unhappy Mind." *Science* 330 (6006): 932.

Kral, T., B. S. Schuyler, J. A. Mumford, M. A. Rosenkranz, A. Lutz, and R. J. Davidson. 2018. "Impact of Short- and Long-Term Mindfulness Meditation Training on Amygdala Reactivity to Emotional Stimuli." *NeuroImage* 181: 301–13.

LeDoux, J. 1996. *The Emotional Brain*. New York: Simon and Schuster.

Lieberman, M. 2014. *Social: Why Our Brains Are Wired to Connect*. New York: Crown.

Litwin, H., E. Schwartz, and N. Damri. 2017. "Cognitively Stimulating Leisure Activity and Subsequent Cognitive Function: A SHARE-Based Analysis." *Gerontologist* 57 (5): 940–48.

Louie, D., K. Brook, and E. Frates. 2016. "The Laughter Prescription: A Tool for Lifestyle Medicine." *American Journal of Lifestyle Medicine* 10 (4): 262–67.

Luders, E., N. Cherbuin, and F. Kurth. 2015. "Forever Young(er): Potential Age-Defying Effects of Long-Term Meditation on

Gray Matter Atrophy." *Frontiers in Psychology* January 21. https://doi.org/10.3389/fpsyg.2014.01551.

Magnuson, C. D., and L. A. Barnett. 2013. "The Playful Advantage: How Playfulness Enhances Coping with Stress." *Leisure Sciences* 35 (2): 129–44.

Manninen, S., L. Tuominen, R. I. Dunbar, T. Karjalainen, J. Hirvonen, E. Arponen, R. Hari, I. P. Jääskeläinen, M. Sams, and L. Nummenmaa. 2017. "Social Laughter Triggers Endogenous Opioid Release in Humans." *The Journal of Neuroscience* 37 (5): 6125–31.

Myrick, J. G., R. Nabi, and N. J. Eng. 2021. "Consuming Memes During the COVID Pandemic: Effects of Memes and Meme Type on COVID-Related Stress and Coping Efficacy." *Psychology of Popular Media*. Advance online publication. https://doi.org/10.1037/ppm0000371.

Proyer, R. 2013. "The Well-Being of Playful Adults: Adult Playfulness, Subjective Well-Being, Physical Well-Being, and the Pursuit of Enjoyable Activities." *European Journal of Humour Research* 1 (1): 84–98.

Salinas, J., A. O'Donnell, D. J. Kojis, M. P. Pase, C. DeCarli, D. M. Rentz, L. F. Berkman, A. Beiser, and S. Seshadri. 2021. "Association of Social Support with Brain Volume and Cognition." *Journal of the American Medical Association Network Open* 4 (8): e2121122.

Schertz, K. E., and M. G. Berman. 2019. "Understanding Nature and Its Cognitive Benefits." *Current Directions in Psychological Science* 28 (5): 496–502.

Singleton, O., B. K. Hölzel, M. Vangel, N. Brach, J. Carmody, and S. W. Lazar. 2014. "Change in Brainstem Gray Matter Concentration Following a Mindfulness-Based Intervention Is Correlated with Improvement in Psychological Well-Being." *Frontiers in Human Neuroscience* 8 (1): 33.

Smith, P. J., J. A. Blumenthal, M. A. Babyak, L. Craighead, K. A. Welsh-Bohmer, J. N. Browndyke, T. A. Strauman, and A. Sherwood. 2010. "Effects of the Dietary Approaches to Stop Hypertension Diet, Exercise, and Caloric Restriction on Neurocognition in Overweight Adults with High Blood Pressure." *Hypertension* 55 (6): 1331–38.

Tonietto, G. N., S. A. Malkoc, R. W. Reczek, and M. I. Norton. 2021. "Viewing Leisure as Wasteful Undermines Enjoyment." *Journal of Experimental Social Psychology* 97: 104198.

Valls-Pedret, C., A. Sala-Vila, M. Serra-Mir, D. Corella, R. de la Torre, M. Á. Martínez-González, E. H. Martínez-Lapiscina, M. Fitó, A. Pérez-Heras, J. Salas-Salvadó, R. Estruch, and E. Ros. 2015. "Mediterranean Diet and Age-Related Cognitive Decline: A Randomized Clinical Trial." *JAMA Internal Medicine* 175 (7): 1094–1103.

Vogel, S., and L. Schwabe. 2016. "Learning and Memory Under Stress: Implications for the Classroom." *Nature Partner Journal Science of Learning* 1: 16011. https://doi.org/10.1038/npjsci learn.2016.11.

Yates, L. A., S. Ziser, A. Spector, and M. Orrell. 2016. "Cognitive Leisure Activities and Future Risk of Cognitive Impairment and Dementia: Systematic Review and Meta-Analysis." *International Psychogeriatrics* 28 (11): 1791–1806.

Yeh, T. S., C. Yuan, A. Ascherio, B. Rosner, W. C. Willett, and D. Blacker. 2021. "Long-Term Dietary Flavonoid Intake and Subjective Cognitive Decline in US Men and Women." *Neurology* 97 (10): e1041–56.

Jill P. Weber, PhD, is a clinical psychologist in private practice in Washington, DC, working with people managing varying degrees of anxiety. She is author of *Be Calm*, writes a blog for *Psychology Today*, and has appeared as a psychology expert in *USA Today*, *The Washington Post*, and on CNN. For more information, visit www.drjill weber.com.

ABOUT US

Founded by psychologist Matthew McKay and Patrick Fanning, New Harbinger has published books that promote wellness in mind, body, and spirit for more than forty-five years.

Our proven-effective self-help books and pioneering workbooks help readers of all ages and backgrounds make positive lifestyle changes, improve mental health and well-being, and achieve meaningful personal growth. In addition, our spirituality books offer profound guidance for deepening awareness and cultivating healing, self-discovery, and fulfillment.

New Harbinger is proud to be an independent and employee-owned company, publishing books that reflect its core values of integrity, innovation, commitment, sustainability, compassion, and trust. Written by leaders in the field and recommended by therapists worldwide, New Harbinger books are practical, reliable, and provide real tools for real change.

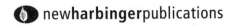

newharbingerpublications

MORE BOOKS from
NEW HARBINGER PUBLICATIONS

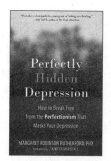
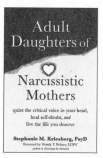

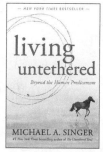